THE ECOLOGY OF ECOLOGISTS

The Ecology of Ecologists

HARNESSING DIVERSE APPROACHES FOR A STRONGER SCIENCE

+ +

JEREMY FOX

THE UNIVERSITY OF CHICAGO PRESS
CHICAGO AND LONDON

The University of Chicago Press, Chicago 60637
The University of Chicago Press, Ltd., London

Published 2026
Printed in the United States of America

35 34 33 32 31 30 29 28 27 26   1 2 3 4 5

ISBN-13: 978-0-226-84495-4 (cloth)
ISBN-13: 978-0-226-84494-7 (paper)
ISBN-13: 978-0-226-84496-1 (ebook)
DOI: https://doi.org/10.7208/chicago/9780226844961.001.0001

Library of Congress Cataloging-in-Publication Data

Names: Fox, Jeremy T. author
Title: The ecology of ecologists : harnessing diverse approaches for a stronger science / Jeremy Fox.
Description: Chicago : The University of Chicago Press, 2026. | Includes bibliographical references and index.
Identifiers: LCCN 2025023508 | ISBN 9780226844954 cloth | ISBN 9780226844947 paperback | ISBN 9780226844961 ebook
Subjects: LCSH: Ecology—Philosophy
Classification: LCC QH541 .F69 2025 | DDC 577—dc23/eng/20250820
LC record available at https://lccn.loc.gov/2025023508

♾ This paper meets the requirements of ANSI/NISO Z39.48-1992 (Permanence of Paper).

Authorized Representative for EU General Product Safety Regulation (GPSR) queries: **Easy Access System Europe**—Mustamäe tee 50, 10621 Tallinn, Estonia, gpsr.requests@easproject.com
Any other queries: https://press.uchicago.edu/press/contact.html

For Brinsley. Thank you for your love and patience.

Contents

Preface ix

Introduction: Ecologists Disagree on What Ecology Is and How to Do It. Good. *1*

1 The Diversity of Ecology *7*

2 The Benefits of Diversity, in Nature and in Ecology *45*

3 Complementarity *51*

4 Selection *66*

5 Diverse Tools for Diverse Jobs: The Many Uses of Mathematical Models *110*

6 Fighting Lack of Diversity: The Value of Contrarians *145*

7 Tying It All Together: The Many Roads to Generality in Ecology *160*

8 The Downsides of Diversity *189*

9 It's Not Just Ecology *199*

10 The Hedgehog and the Fox *209*

Acknowledgments 221
References 223
Index 271

Preface

What This Book Is About

This book is about how ecological research can better harness the diversity of goals, ideas, and approaches of ecological researchers. But it is not a textbook or an instruction manual. Rather, it's a guidebook.

Ecology is a broad and heterogeneous field, so every ecologist needs to decide what questions to ask and how to answer them, and then explain and justify those choices to others. We all get plenty of practice at this, as authors and reviewers of research proposals, grant proposals, and scientific papers. But that practice tends to be narrowly focused on the particular proposal or paper at hand. The premise of this book is that you can learn a lot about what questions to ask and how to answer them not just by thinking about your own work, or the work of others on the same topic, but by thinking about work on other topics as well, and even work from different scholarly fields. This book will help you do that. It is a guidebook to the diverse ways in which ecological research is done, or could be done.

Who This Book Is for and How to Read It

This book is aimed at graduate students and postdoctoral researchers in fundamental ecology (I don't know enough about applied work to write a book about it). I think that established researchers will find the book interesting as well, though they may wish to skim discussions of topics and examples with which they are already familiar. I am less sure of the value of the book to advanced undergraduates. My goal is to prompt critical thought, which requires the reader to have a foundation of background knowledge and opinions against which to evaluate the book's arguments.

You can read this book on your own, but I think and hope it is best read as part of a graduate seminar or reading group. It's meant as fodder for discussion and debate.

Some Other Things You Should Read besides This Book

Like any good guidebook, this one is personal and opinionated. Think of ecology as a big city, and me as a local who's lived there his entire life and likes showing people around.

Like any tour guide, I'm just one person. I have my own experiences and perspectives. Some of which are more or less widely shared with those of other ecologists, others of which are more or less unique to me. If you were touring the field of ecology with a different guide, you'd get a different tour. So don't take anything I say as gospel, or assume it's representative of all ecologists, or even of all ecologists who have some things in common with me. And definitely don't just read what I have to say! You'll be a better ecologist, and you'll also get more out of this book, if you read a bunch of other stuff too. Here's a short and inevitably *very* incomplete list of suggestions, in no particular order. Note that I haven't read everything on this list (yet!). In some cases I'm sharing recommendations from colleagues, blog commenters, and book reviewers.

- I started grad school in 1995, just as the field of ecology was starting to change quite fast in some quite big ways (see chapter 1). Ecologists as a whole are asking different questions now. The concepts and topics that I was taught to think of as the "core" of ecology aren't really the "core" of the field any longer. Which raises the question of whether they're still worth reading about, and if so, why. Personally, I think they're still worth reading about, for various reasons. I think many of those former "core" topics remain scientifically interesting, even though most ecologists these days study other topics. And I think knowing something about the history of your field helps you do better science in the present day, even if you don't study those former "core" topics. If you want to learn about what were once the "core" topics of ecology, you can't go wrong with *Foundations of Ecology: Classic Papers with Commentaries*, and *Foundations of Ecology II*. Does what it says on the tin, as the British say. Each chapter lays out the history of ecological research on some "core" topic. Each chapter comprises a reprint of a foundational paper on the topic, followed by a historical review of subsequent research. The first volume covers earlier classic papers, the second volume covers papers from 1970 to 1995. Other books in the same series use the same format to cover various subfields of ecology: *Foundations of Tropical Forest Biology*, *Foundations of Biogeography*, and *Foundations of Macroecology*.

- Frank Egerton has a long series of articles on the history of ecological sciences in the *Ecological Society of America Bulletin*. It starts in ancient Greece and works its way forward to the year 2000. The focus is largely on work in Europe and North America. I've dipped in and out of the series. The series is available for free online: https://esajournals.onlinelibrary.wiley.com/doi/toc/10.1002/(ISSN)2327-6096(CAT)Collections(VI)CollectionsBulletin.
- Sharon Kingsland is a leading historian of ecology. Her *Modeling Nature* is an outstanding history of population ecology up until the 1960s and '70s. You'll learn a lot about how mathematical models are viewed in ecology by reading *Modeling Nature*. Sharon Kingsland also wrote *The Evolution of American Ecology, 1890–2000* and recently published *A Lab for All Seasons: The Laboratory Revolution in Modern Botany and the Rise of Physiological Plant Ecology*.
- Laura Martin's *Wild by Design: The Rise of Ecological Restoration* is a history of an increasingly important area of applied ecology.
- The last decade or so has seen a rapidly growing number of papers and books that reframe, rethink, or critique what one might think of as "conventional" histories of ecology. Some of these papers and books foreground the achievements and perspectives of people who often are omitted from conventional histories of ecology (e.g., women; ecologists writing in languages other than English). Others discuss widespread, long-standing biases in researchers' thinking on various topics. Histories of natural historical, ecological, and other scientific work in Latin America include Pablo Gómez's *The*

Experiential Caribbean and T. P. Glick's article, "History of Science in Latin America," in *The Cambridge Encyclopedia of Latin America and the Caribbean*. Abena Dove Osseo-Asare's *Bitter Roots* is a classic history of botany in Africa, especially in relation to medicinal plants. Ambika Kamath and Melina Packer's *Feminism in the Wild* provides a feminist critique of behavioral ecology. In 2022 the journal *American Naturalist* published a collection of papers from scientists and social scientists, arguing that "systems of oppression" have shaped the content and practice of ecology and adjacent fields (Kamath, Velocci, et al. 2022 and references therein).

- I enjoy reading biographies and memoirs of scientists, but I confess that I don't usually take any inspiration or lessons from them. Scientists who merit a biography or memoir tend to be too singular for that. The one exception for me is population ecologist Dennis Chitty's memoir, *Do Lemmings Commit Suicide? Beautiful Hypotheses and Ugly Facts*. It's an unusual scientific memoir in that it's a record of admitted failure. Chitty spent his entire career trying and failing to explain population cycles in small mammals. It's a good book for thinking about the determinants of progress (or the lack thereof) in ecological research. Chitty is unusually explicit about the philosophy of science that guided his choices as to what studies to conduct and how to interpret their results. But your mileage may vary. Just because I mostly don't find scientific memoirs inspirational or instructive doesn't mean you won't. Other memoirs and biographies of ecologists, and scientists working at the interface of ecology and adjacent fields, include but aren't limited to E. O. Wilson's memoir *Naturalist*, Nancy Slack's G. E. Hutchinson biography, *G. Evelyn Hutchinson and the Invention of Modern Ecology*, Hope Jahren's memoir *Lab Girl*, and Jasmine Graham's memoir *Sharks Don't Sink: Adventures of a Rogue Shark Scientist*.
- In contrast to memoirs and biographies of scientists, I've found it quite helpful to read some philosophy of science. Even good scientists often struggle to articulate what makes for good science—what makes for a good scientific question, or a good answer to a scientific question. Philosophers of science sometimes are better at articulating what makes for good science than scientists themselves are. Michael Strevens's *The Knowledge*

Machine is an unusual and excellent philosophy of science book. It's a popular rather than an academic book, so it's very readable and accessible. And it presents Strevens's own ideas, in addition to summarizing the work of other philosophers of science. Peter Godfrey-Smith's *Theory and Reality* is a more conventional introductory treatment of the major lines of thought in philosophy of science. If you're looking for something more specific to ecology, ecologists Mark Taper and Subhash Lele's edited volume *The Nature of Scientific Evidence* is worth a look. Some of the chapters in Resetarits and Bernardo's edited volume *Experimental Ecology* also are worth a look, to see the contrasting ways in which different ecologists think about what makes their own research programs worthwhile.

- "Biodiversity" might be the single most important word in ecology today. Back when I started grad school, the term had only just been coined. The story of its coinage is quite interesting, and I think quite important for any ecologist today to know. David Takacs's *The Idea of Biodiversity: Philosophies of Paradise* chronicles the history of the term "biodiversity" through interviews with those who coined the term and promoted its use early on.

Introduction: Ecologists Disagree on What Ecology Is and How to Do It. Good.

Ecology has long been plagued by two linked anxieties: that we don't know what it is, and we don't know how to do it.

It may seem preposterous to claim that ecologists don't know what ecology is. Open any ecology textbook, and on the first page you'll see ecology defined as "the study of living organisms, and their interactions with one another and their environment," or words to that effect. Except that, when you read the rest of the textbook to discover what that means, you'll find a heterogeneous mix of material that puzzles many newcomers to the field (Vellend 2016). How do chapters on levels of hierarchical organization (individual organisms, populations, communities), abstract concepts like food webs, and topics as different as evolution, competition, the nitrogen cycle, climate change, and conservation all fit together into a unified whole? The answer might well be that they don't. Philosopher of science Gregory Cooper (2003) devoted an entire book to searching for a coherent definition of "ecology"—and settled on a definition that, by his own admission, excluded entire subfields! And in an unscientific 2017 poll of 232 readers of the *Dynamic Ecology* blog, the readership of which is composed largely of ecologists and ecology graduate students, a 44% plurality characterized the field of ecology as a "disunified mess" or something close to it (J. Fox 2017).

To claim that ecologists don't know how to do ecology might seem equally preposterous. And it is a bit imprecise. Actually, all ecologists think they know how to do ecology—and that it's only other ecologists who don't. At least, that's the impression one gets from reading what prominent ecologists have said about how to do ecology. Ecology in the

1960s either made a great leap forward under the influence of Robert MacArthur and his emphasis on discovering and explaining general patterns (J. Brown 1997), or it made a great leap forward in the 1980s and '90s once it quit focusing on static patterns and started worrying about dynamics (Kareiva 1997). Testing alternative hypotheses in ecology either is possible only rarely (Quinn and Dunham 1983), or often (Loehle 1987). Small-scale field experiments either reveal the mechanisms driving ecological phenomena (R. Paine 2010), or are of limited use because all the interesting and important phenomena happen on larger spatial scales (J. Brown and Maurer 1989). Ecology's increasing emphasis on mathematics and statistics has either increased the field's rigor and allowed ecologists to ask interesting new questions (Caswell 1988; Scheiner and Willig 2011; Evans et al. 2013; Marquet et al. 2014), or has set ecology back by unmooring it from its foundation in natural history (Dayton and Sala 2001; Simberloff 1981). Meta-analysis either adds immense value by providing an objective statistical summary of the primary literature (Koricheva, Gurevitch, and Mengersen 2013), or incentivizes ecologists to engage in question-free data dredging, and to parasitize other people's data rather than collecting their own (Lindenmayer and Likens 2011, 2013). Experiments in artificial systems such as laboratory microcosms either allow ecologists to obtain valuable data that couldn't be obtained any other way (Gause 2019; Lawton 1996; J. Drake and Kramer 2012; J. Fox 2011b), or are too artificial to teach us anything about how nature works (Carpenter 1996; Krebs 2015b). And no, I'm not just cherry-picking disagreements among the rare ecologists willing to publicly criticize other ecologists' research. In an anonymous 2019 poll on the *Dynamic Ecology* blog, almost everyone agreed that ecological research suffers from numerous serious problems—but no one agreed on what those problems are (J. Fox 2019b).

Both anxieties reflect ecology's development from disparate sources (Kingsland 1995, 2005). Ecologists differ from one another in their personal motivations and scientific goals, and they always have. Even if you just focus very narrowly on those traditionally regarded as the founders of ecology in the English-speaking world—Tansley, Cowles, Shelford, Forbes, Lotka, Volterra, Gause, Clements, Gleason, Elton, and others—you'll be struck by their differences in terms of the questions they asked and how they went about answering them. Those profound differences in motivation, methods, and goals are still with us today.

Proposals to address these two anxieties tend to run to type. One type argues that ecology is, or should be, about discovering universal laws of nature, analogous to the laws of physics. Or if not universal laws, then theoretical principles and empirical regularities that play much the same scientific role as universal laws. Failing to discover the generalities hidden beneath the apparent polyglot variety of living nature would condemn ecology to be nothing but a collection of unrelated case studies. Proposals of this type include MacArthur (1972), J. Brown (1997), Lawton (1999), B. Murray (2000), Turchin (2001), Berryman (2003), Colyvan and Ginzburg (2003), Scheiner and Willig (2008), Dodds (2009), and Harte (2014).

A second type of proposal inverts the first, arguing that there are no universal laws or other important generalities to be discovered in ecology, and that attempts to discover them are useless at best. Rather, ecology is a science of unrelated case studies, and that's a good thing. Successful case studies allow ecologists to predict and control the behavior of specific ecological systems. This is essential if ecology is to comprise more than just empty word games and ungrounded speculation, and if ecology is to be useful for conservation and management. The generality in ecology concerns not ecology itself, but how ecologists should do ecology. There are universally applicable methods that ecologists can use to study any particular system or address any particular management problem. R. Peters (1991), Shrader-Frechette and McCoy (1993), Kareiva (1997), and Simberloff (2004) are proposals of this type.

My thesis is that both types of proposal are wrong—and right.

The mere fact that those two anxieties—about what ecology is and how to do it—exist at all make it hard to imagine that they could be resolved by either type of proposal. Proposals that ecology is about the search for generalities, and proposals that it's about applying universally applicable methods to specific situations, share the assumptions that ecology is one thing, and that there is one right way to do it. But if there really was one Correct Definition of "ecology" and one Right Way to do it, we'd likely have agreed on it by now. After all, consider the alternative: that there is one Correct Definition of ecology and one

Right Way to do it, but for some reason we can't agree on it even though we've been doing ecology for well over a century. I find that alternative both too implausible and too depressing to contemplate. It suggests that ecology as a field is deeply and permanently dysfunctional, despite being pursued by extremely smart and hardworking ecologists. In this book, I will instead argue for a far more plausible and happier possibility: that the existence of ongoing disagreement about what ecology is and how to do it suggests that both sides have a point. Both are right in some respects and in some circumstances, and wrong in other respects and circumstances.

Indeed, I will go one step further and argue that disagreement among ecologists about what ecology is and how to do it is not a sign of the weakness of ecology, but a sign of strength. In my view, the diversity of ecologists' motivations, goals, expertise, and approaches gives the field a collective vigor it would otherwise lack. A diverse field of ecologists can solve a wider range of problems than any single type of ecologist could, and can solve problems that would be intractable to any single type of ecologist, or even any single collaborative research group. Ecologists know that no single living species can be well adapted to all environments or perform all ecosystem functions and processes on which life depends. That's why diverse groups of organisms often will function better than less diverse groups. The same is true of ecologists themselves. There isn't just one sort of ecology that's worth doing. There are several (at least), each of which can best be pursued via a characteristic mixture of complementary approaches. Which is why it's a good thing that ecology has always been done by ecologists as different as Gause and Clements. Otherwise, much ecology that's worth doing would go undone.

I suspect that, for many readers, I'm preaching to the converted here. Disagreements about what ecology is and how to do it seem to be largely a thing of the past. As illustrated by the fact that all the disagreements I cited above are at least a decade old. Ecologists these days mostly take it for granted that diversity of goals, ideas, and approaches is a good thing. Does that mean we all now agree on what ecology is and how to do it? I don't think so. Recall those recent poll results: ecologists today see the field as a disunified mess. And they all agree the field has some serious problems, while disagreeing on what those problems are. Put it all together, and what emerges is a picture of a highly fragmented field, comprising highly specialized researchers (or, increasingly, specialized collaborative research groups). We're all narrowly focused on our own research programs, and we're all well aware of the problems those programs face. We're all glad that there are lots of other ecologists, and other collaborative groups, out there with their own research programs. But we're glad about it in the same abstract way that we're glad about the existence of faraway species most of us will never see in the wild. We're only vaguely aware of the details of those other ecologists' research programs and the problems those programs face.

Don't get me wrong: a field that's diverse in the sense of being highly fragmented definitely has its strengths. Ecologists today ask a greater range of questions, in more places, than we ever have before. We also collaborate with one another more often and more extensively than we ever have. But we're also missing a lot of opportunities to learn from one another if we all just keep our heads down and beaver away on our own questions, in our own places. Even if we also collaborate with others who are addressing the same questions in other places. There's more to harnessing diversity of goals, ideas, and approaches in ecology than just "asking more questions in more places," or "collaborating with more people." In this book, I'll try to spell out why that is.

Spelling out why that is sometimes will require being critical, because ecology is hard. Smart and hardworking as ecologists are, none of us are infallible. Variation among individuals is essential if natural selection is to adapt a population to current environmental conditions, or new conditions. Analogously, variation among ecologists and their approaches is essential if ecologists are to select effective approaches for any given question. But that very same variation also makes it challenging to select the most effective approach(es) for any given question. So there's never any shame in ecologists asking, "How are we doing, and could we be doing better?"

That question lacks a single, simple answer. There is no single, infallible, step-by-step recipe for Doing Good Ecology. But neither is it a case of anything goes. There may not be hard-and-fast rules, but I will argue that there are context-dependent rules of thumb that can be identified via a comparative approach. Researchers working on different ecological topics often use similar research approaches. Each research approach has its own characteristic strengths, limitations, and "failure modes" (ways in which it tends to go wrong, when it does go wrong). It is inefficient for ecologists working on different topics to independently reinvent the wheel—and independently rediscover the circumstances in which the wheel is likely to get stuck. Throughout the book, I use a comparative approach to illustrate the characteristic strengths, limitations, and failure modes of different research approaches. The first goal is to help ecologists do more of what works and less of what doesn't. The second goal is to help ecologists recognize that "what works" isn't one thing, because ecology isn't one thing. Yes, our field is a disunified mess—but not in any way that we could ever clean up, or would ever want to clean up. It's time we embraced the polyglot variety of ecology as our field's greatest strength.

1 *The Diversity of Ecology*

Diversity of Ecologists versus Diversity of Ecology

Recent writing about diversity in ecology focuses on the diversity of ecologists, particularly in terms of attributes such as gender, race, ethnicity, sexual orientation, and socioeconomic background (e.g., Shaw and Stanton 2012; Yoder and Mattheis 2016; Cid and Bowser 2015; Shaw, Accolla, et al. 2021; Kamath, Velocci et al. 2022; Beck et al. 2014; Primack et al. 2023; Woods, Leggett, and Miriti 2023). Certainly, the narrow subset of ecologists with which I happen to be most familiar—academic ecologists and graduate students in North America—is now much less male-dominated, and much less white, than it was even when I started graduate school in 1995, never mind back in the early decades of the 20th century, when the first scientific societies of ecologists were founded. One can of course note this progress while also noting that academic ecology in North America still has a ways to go in terms of achieving a diverse, representative field in which everyone feels valued (e.g., Primack et al. 2023).

But important as the diversity of ecologists in every dimension is—and it's very important!—it's not what this book is about. So I'm now going to set that to one side, not because diversity of ecologists is unconnected to the diversity of our ideas and approaches, but because the connections are too complex, too rich, and too personal to generalize about. By way of illustration, here are stories of how I, and some other ecologists, got into ecology.

How I Got into Ecology

I'm a strange ecologist.

OK, that's an exaggeration. In many respects, I'm not strange at all. For starters, I'm a white man employed as a professor by a big North American research university. I don't stand out from the crowd at the Ecological Society of America annual meeting. I'm very lucky that I've never sat in a room full of other ecologists and felt like I didn't belong.

In other respects, though, I'm quite different from many other ecologists. Awareness of these differences made me a bit self-conscious and defensive about my own research for many years. Don't misunderstand—I love ecology, I love being an ecologist, and I always have. But for a long time, I felt like a bit of an outsider. I felt that way not because of my appearance or my job, but because of my motivations and research approach. It wasn't a big deal; most days I didn't feel any outsider-y feelings at all. And I certainly wouldn't equate my own occasional mild outsider-y feelings with those of someone who has had the experience of (say) being the only black person, or only woman, or only black woman, in a room full of ecologists. But I think it's only human to be self-conscious about the things that make you different from those around you, even if those things aren't really *that* big a deal, or are a consequence of your own choices.

I didn't get into ecology because of a love of wild nature, or out of a desire to conserve endangered species, or to work with my favorite animals. Again, don't misunderstand—I enjoy being out in nature, I support conservation, and I think sharks are awesome. In high school as well as most of the way through college, I thought I wanted to be a rocky intertidal ecologist. I thought that because I liked poking around in tidepools on jetties at the beach in New Jersey as a kid. But really, none of that is why I became an ecologist. Here's why I became an ecologist (J. Fox 2016a):

> Looking back, it was only after becoming an ecologist that I fully understood what really attracted me to ecology. It wasn't love of the outdoors or love of natural history or a desire to save the earth, at least not mostly. I liked those things well enough—well, except for natural history, which seemed to me like a lot of rote memorization

> and for which I had no aptitude. . . . [W]hat attracts me to ecology is the conceptual side of the field. Besides ecology, my favorite subject in college was philosophy. I like being in a field in which what questions to ask, and how to answer them, are somewhat up for grabs (without being so up for grabs that there's little possibility of progress, as in some fields of humanities and social sciences). Those sorts of foundational and methodological issues are borderline philosophical issues. And I like being in a field in which one can connect abstract ideas to real life.

And so I didn't end up working in the rocky intertidal. Instead, I became an indoor ecologist. I've spent my whole career working in protist microcosms: glass bottles containing tiny artificial ponds. In these microcosms grow communities of single-celled protists such as *Paramecium*, along with bacteria and other microscopic organisms. These communities play by the same basic rules as natural communities. The challenge, as in nature, is to figure out what those rules are. I find satisfaction in figuring out those rules, rather like the satisfaction of solving a puzzle or a riddle. Success is essential to satisfaction—valiant but ultimately unsuccessful attempts at puzzle-solving aren't satisfying. Which is why I work in microcosms. Figuring out the rules that ecological communities play by is rather easier in microcosms than in many natural systems. Protists reproduce very fast, so you can track fluctuations in their abundances for hundreds of generations in just a few months. Microcosms are controllable, manipulable, and replicable. And there aren't any ethical issues involved in experimenting on nonpathogenic protists. So you can get unique data from microcosms that you just couldn't get otherwise. In this way, microcosms complement field studies. My reasons for working in protist microcosms are more or less the same ones as Russian ecologist G. F. Gause back in the 1930s, when he conducted his now-classic microcosm experiments on predation and competition. Gause is perhaps the most famous protist microcosm researcher in ecological history, but he wasn't the first or last (Jessup, Forde, and Bohannan 2005). Since Gause's day, there've always been a few ecologists working in protist microcosms.

You might think the lengthy history of protist microcosms in ecology would imply that microcosms are as widely accepted (if not as widely used) as, say, fruit flies are in genetics research. But you'd be wrong. I was a second-year grad student back in 1996 when I opened the latest issue of the journal *Ecology* to find a paper by no less than Steve Carpenter (1996)—one of the world's leading ecologists at the time—arguing that all microcosm work was worse than useless. Not just uninformative, but positively misleading. Carpenter even claimed that microcosm research was a threat to the future of ecology: a whole generation of students who would otherwise go on to become good field ecologists might instead fall for the siren song of microcosms, leaving too few people capable of doing field research. I soon learned that Carpenter was far from alone in his views (reviewed in Jessup, Forde, and Bohannan 2005). No less than G. E. Hutchinson, a key founder of modern ecology, criticized the sorts of microcosm experiments I do. Hutchinson (1978) refers to a microcosm as "a highly inaccurate analogue computer . . . using organisms as the moving parts" (quoted in Jessup, Forde, and Bohannan 2005, 286). Hutchinson's student Robert MacArthur, another founder of modern ecology, dismissed what he called "bottle experiments" as uninteresting and unimportant (Jessup, Forde, and Bohannan 2005, 286). MacArthur's negative opinion likely contributed to the decline in the use of microcosm experiments in ecology in the 1960s and '70s (Mertz and McCauley 1980; Ives, Foufopoulos, et al. 1996).

So from the time I started grad school, I felt I had to get very good at defending myself. My papers routinely included blanket defenses of microcosms as a research approach—the sort of blanket defense that no ecologist who does fieldwork ever feels obliged to make. And when I started blogging about ecology, one of my first posts was titled "Objections to Microcosms in Ecology, and Their Answers" (J. Fox 2011b). A part of me actually enjoyed this; I like a good debate. But part of me was frustrated and annoyed, because it felt like a bad debate, not a good one. It felt pointless, taking one side of the same old arguments ecologists had been having for decades. But I kept at it, because deep down I felt like I had no choice. I felt that if I couldn't convince enough people that protist microcosms were a useful research tool for ecologists to have in their toolbox, then I wouldn't have an academic career and I'd have to go do something else with my life (J. Fox 2011a). There wouldn't be a place for someone like me in ecological research if I didn't fight for it.

In retrospect, I probably overestimated the prevalence of anti-microcosm views in ecology. I definitely overestimated the risks that those views posed to my career (J. Fox 2016b). But my feelings at the time were understandable.

Any ecologist could tell a story like this. We're each unusual compared to other ecologists, in some respects, including the ways we got into ecology in the first place. And those diverse ways of getting into ecology don't always map very neatly onto the research we end up doing.

The Diverse Ways Ecologists Get into Ecology

There's a stereotype that ecologists all start out as nature lovers. Certainly, many do. Here's how ecosystem ecologist and former Ecological Society of America President Steward Pickett got into ecology (Sprugel 2016b):

> As a child, Steward spent summers in Louisville, Kentucky, at a Boy Scout camp run by his father. Steward described himself as a "free agent" at the camp, able to explore the woods by himself. "I didn't really know what I was seeing, but I knew it was neat. This coincided with my mom bringing home a book on ecology. That solidified it; I knew what I wanted to do."

Here's avian conservation biologist Auriel Fournier (2016):

> There are pictures of me geeking out about birds since I was quite small. Those who knew me even in passing as a kid are unsurprised where I ended up. . . . I grew up out in rural Ohio on my Opa's (grandfather in Dutch) farm. It was a perfect place to grow up, 100 acres of fields, stream and forest to explore, play on and explore.

And here's my University of Calgary colleague, wildlife conservation biologist Allyson Menzies (2020):

> I am of mixed Red River Métis and Settler descent, born and raised in Treaty 1 & 2 territory and the homeland of the Métis Nation (a.k.a Manitoba). My curiosity about and love for the natural world is

> rooted in a childhood spent admiring prairie sunsets, summer thunderstorms, northern lights, and many camping adventures with my family. . . . Devoting my professional life to environmental science and wildlife conservation is no coincidence; it allows me to feel connected to my Métis heritage, and the ancestral right and responsibility to care for the land.

But sometimes the nature bug doesn't bite you until you're much older. Here's ecologist Jean Richardson (Sprugel 2016a):

> Unlike many ecologists I know, I do not think I can claim any interest in ecology until my university days. Student projects introduced me to the world of research and I loved it so much, I just kept looking for ways to [keep] doing it!

The nature bug can bite you even if you're not out in "nature." Here's pollination ecologist and professional naturalist Jeff Ollerton (2015):

> The grassland in which [my childhood friends and I erected] a tee-pee . . . is not some country meadow, the kind of wild rural landscape cited by so many other naturalists as inspiring their childhood fascination with natural history. These grasslands had arisen spontaneously on cleared demolition sites, following the removal of Victorian terraced housing and tenement blocks, some of which were slums and others that had suffered bomb damage in the Second World War. . . . So you don't have to have had a rural upbringing to appreciate and benefit from nature, and to later influence your profession and passions, any piece of land can inspire interest in kids, regardless of its origin, if nature is left to colonise.

Finally, the nature bug needn't even bite you at all. As I said above, it didn't bite me; it didn't bite quantitative ecologist Ben Bolker (2005, 550) either:

> Quantitative ecologists are only loosely anchored by the natural history of particular systems. Even the word "systems" is a giveaway; we

> see organisms as realizations of ideas, not as furry, feathery, or green individuals. Many of us came to ecology from physics, or mathematics, or statistics, because we loved its ideas. If we didn't care about the organisms, we would have been content as mathematicians or physicists, but our true love was for the way that real ecological communities could embody general mathematical concepts of dynamics and variation.

Nor did the nature bug bite this anonymous black woman ecologist who got into ecology in part because she had to. Her experiences growing up didn't enamor her of outdoor environments or make her comfortable in them. But she needed to get comfortable for the sake of her students (Woods, Leggett, and Miriti 2023, 335):

> Prior to pursuing a PhD in ecology, my interest in ecology stemmed from having to be outside. I was teaching at a liberal arts college in the Southeast US and the faculty had a goal of teaching our students ecology in a natural environment. Like many underrepresented minorities, my experiences in the natural environment were few. Growing up, I did not hike, go camping, or visit national parks. . . . I was never exposed to ecology, and I was not comfortable in outdoor environments. However, I became comfortable in nature because I felt the need to be an effective leader for the students I taught and supported the institution's goal of increasing the field experience of students learning the biology curriculum.

And even if the nature bug did bite you, well, are you sure that's why you became an ecologist? I think there's a lot of wisdom in forest ecologist Marcus Eichhorn's (2016) remarks on why he became an ecologist, in a piece aptly titled "A Lie about My Childhood":

> Anyone involved in admissions or graduate recruitment in ecology will be familiar with the stereotypical opening of the personal statement:
>
> "When I was a child, I loved to play outside in nature. I watched the birds and the insects and the flowers and I knew that I wanted to spend my life studying them."

Something along these lines opens the majority of the applications I read each year. Perhaps for some it's actually true, though I suspect that most are teleological. Either the author is trying to convince me, or has already convinced themselves, that the whole direction of their life has been moving steadily and inexorably towards ecological research from their very first awakenings of consciousness. Who am I, hard-hearted cynic, to stand in the way of manifest destiny?

Why am I so sceptical? I too am passionate about nature. I genuinely love being outdoors, collecting data, or simply observing natural systems and trying to figure out how they work. I grew up in the countryside and was most at peace when taking my dog for long walks through the fields and woodland or climbing trees. This bucolic upbringing is bound to have had a lasting influence on my chosen direction in life.

And yet . . . the evidence for a similar effect isn't there from anyone else in the village, other than those who have continued on the family farm, for whom options were more limited. My siblings and friends from childhood include a doctor, dinner lady, teacher, policeman . . .

There are other anecdotes I could pull together to tell a partial story. We did tie old wellies to a rope and throw them in a pond to try and catch newts. That probably happened a handful of times and I don't recall ever reeling in anything but mud. I remember my father encouraging me to help in the vegetable garden, and the excitement at eating my first crop of radishes. They were to be my only harvest, and any further assistance was through compulsion. It may be true that I once took myself off into the woods in Germany, disappearing for a whole day to the great consternation of my parents, then casually strolling back into town at dusk as the search parties were being assembled. I wasn't lost in the embrace of nature; I just wanted to get away from the family for a bit.

I could tell a different story, of the boy who came home from school every evening and promptly ran upstairs to play Sensible Soccer on his Amiga until his hands developed callouses. The child who lagged behind on family walks bleating about the imposition. A bookworm, happier sat indoors reading science fiction than out in the sunshine. All these would be equally accurate, if similarly selective.

Eichhorn goes on to say that he went into ecology in part because of the opportunity for travel. Not only is this fine, it's just as fine as any other reason. Ecologists get into ecology for all sorts of reasons. Those reasons vary from one ecologist to the next, often aren't clear even in retrospect, and often involve a large element of chance. Finally, our diverse reasons for getting into ecology don't always reveal much about our diverse research goals, or our diverse ways of achieving those goals. The same could be said for the diversity of ecologists on other dimensions: this kind of diversity is vitally important, but not because it reveals much about the diversity of our research goals and approaches.

Which is why I'm going to set the diversity of ecologists aside and turn to the focus of this book: the diversity of ecologists' research goals and approaches.

Diversity of Goals and Approaches

So what about diversity of ecologists' research, as opposed to diversity of ecologists? What are some of the biggest ways ecological research is changing, in terms of ecologists' research questions or goals, and the ways we try to answer those questions and achieve those goals? Is ecological research changing in such a way as to diversify? Or are some research goals and approaches replacing others, so that the mix of goals and approaches is changing but their diversity isn't? Are there any ways ecological research is *not* changing? Finally, why has ecological research changed, or not changed?

BIG CHANGES

Let's first review some of the biggest changes in ecologists' research goals and approaches. I'll focus primarily but not exclusively on changes since the late 1980s. That's when the single biggest change began; most of the other changes on which I'll focus started after that.

Big Change 1: Research Focused on Global Change and Other Applications

Perhaps the single biggest change in ecological research since I started grad school in 1995 (or really, since the late 1980s) is that it's now much more focused on applied topics: global change, conservation, invasive species, ecosystem services, etc. Anderson et al. (2021) quantified the frequency of 1-grams and 2-grams (one- and two-word terms) in the full text of papers from 52 ecology and conservation journals over an 80-year period ending in 2010. The majority of the 1- and 2-grams that increased most rapidly in usage in the 1990s and 2000s were associated with global climate change and other applications (fig. 1.1).

The rise was so rapid that papers on these topics went from a negligibly small portion of the ecological literature to a plurality in fewer than 20 years. Papers on just four broad applied topics (global change biology, conservation biology, invasion biology, and restoration ecology) contributed just 1.8% of ecology papers published in 1990, but

Figure 1.1. Fastest-increasing 1-grams and 2-grams (single-word and two-word terms) in 52 ecology and conservation journals during the 2000s

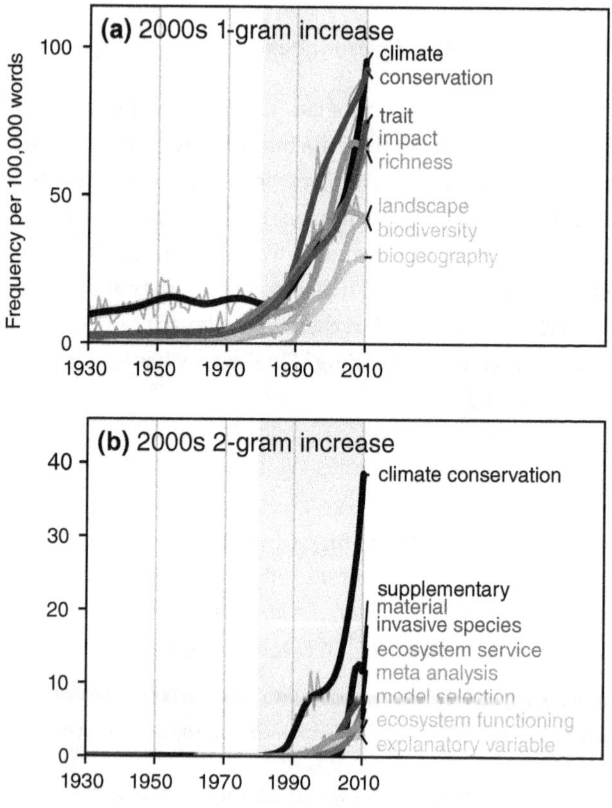

Source: Modified from Anderson et al. (2021).

22.9% of ecology papers published in 2017 (Staples et al. 2019). The same rise shows up if you restrict attention to the most-cited ecology papers. Descriptive studies of plants and wildlife dominate the most-cited ecology papers published before 1980. These days, though, the most-cited ecology papers concern human impacts on nature (Zettlemoyer et al. 2023).

You also can see this collective shift toward applied research since the late 1980s in the founding of new journals, new organizations of ecologists, and new long-term research programs. In terms of new journals, *Conservation Biology* began publishing in 1987, *Ecological Applications* in 1991, *Restoration Ecology* in 1993, *Global Change Biology* in 1995, *Urban Ecosystems* in 1997, the *Journal of Applied Ecology* and *Ecology and Society* both in 1998, *Biological Invasions* in 1999, *Frontiers in Ecology and the Environment* in 2003, *Ecosystems and People* in 2005, the *Journal of Urban Ecology* in 2015, *Ecological Solutions and Evidence* in 2020, and *Earth Stewardship* in 2023.

In terms of new organizations of ecologists, the Society for Conservation Biology was founded in 1985. The Society for Ecological Restoration was founded in 1988. Existing ecological organizations also reorganized themselves to give more prominence to applied research. In 2003, the Association for Tropical Biology changed its name to the Association for Tropical Biology and Conservation. Ecological Society of America (ESA) members can organize themselves into official sections around their shared interests. Most of the sections founded since 2000 that concern ecological topics (rather than, e.g., attributes of ecologists) concern applied or otherwise human-related topics (Agroecology Section, Rangeland Ecology Section, Policy Section, Human Ecology Section, Climate Adaptation Section, Traditional Ecological Knowledge Section, Urban Ecosystem Ecology Section, Invasion Ecology Section, Environmental Justice Section). This is so, even though the ESA has had an Applied Ecology Section since 1971!

In terms of long-term research programs, the US National Science Foundation started funding Long Term Ecological Research (LTER) sites in urban areas in 1997. It began with the Baltimore Ecosystem Study and the Central Arizona–Phoenix LTER site.

Now, these examples of new journals, organizations, and long-term research programs are all anecdotal. They're all examples I happen to know about, not examples I identified via a systematic search. But they align with the systematic text mining and bibliometric data discussed above, which suggests that they're anecdotal symptoms of a real trend.

Indeed, the shift toward applied topics is so strong and universal that you can even detect it within applied subfields of ecology. Applied ecological research is becoming even *more* applied. The authors of Hintzen et al. (2020) used text mining of 32,000 articles published in 16 ecology and conservation journals from 2000 to 2014 to show that conservation biology is increasingly human-focused. Conservation biology papers increasingly focus on human aspects of conservation, such as sociology and politics, rather than ecological aspects. Citations of "purely" ecological papers and topics in conservation journals are declining.

Do these shifts represent diversification of ecologists' research interests? Or has research on global change, conservation, and other applied topics replaced research on older topics? No doubt it's some of both; the two possibilities aren't mutually exclusive. It could also be one followed by the other. When this shift first started in the late 1980s, it represented a diversification of ecologists' collective research interests. But I think the shift has proceeded to the point where it's now best described as primarily a replacement of fundamental research by applied research. The authors of McCallen et al. (2019, 109) used machine learning to show that papers on "anthropogenic themes," as well as "data-intensive" papers, have replaced papers on older theoretical topics. Other analytical approaches give similar results. Staples et al. (2019) finds that papers on global change, invasion biology, conservation biology, and restoration ecology rarely mention terms commonly associated with fundamental ecological theories in their titles or abstracts. That is, a growing and now-substantial fraction of all ecology papers come from subfields that don't even refer in passing to older and more-theoretical topics. The increasing frequency of ecology papers on anthropogenic impacts and climate change is disproportionately concentrated in high-impact journals (Knott et al. 2019). The authors of Craven et al. (2019) used natural language processing and network analysis to demonstrate declining conceptual diversity in papers on "biodiversity." But the indicator of replacement that resonates most with me personally comes from Anderson et al. (2021). The authors found that mentions of the 1-grams "competition" and "predation" peaked in the late 1980s and then plummeted. My own doctoral research, conducted in the late 1990s and published in the 2000s, set out to test theoretical models of competition, predation, and their interplay (J. Fox 2002, 2007). At the time, I thought of myself as working on a topic at the core of the field of ecology. Apparently, I was wrong.

Big Change 2: More-Advanced Statistics

Another big change in ecological research over the past few decades is the rise of more-advanced statistics. In the decades before I started graduate school in 1995, by far the most common statistical methods in ecology were linear regression and ANOVA: linear models with no hierarchical structure, and no random effects besides normally distributed sampling error. Model comparison was limited to frequentist statistical testing of null hypotheses. All that was just starting to change in 1995, and by the time I finished my postdoc, in 2004, the changes had accelerated noticeably (Touchon and McCoy 2016; Low-Décarie, Chivers, and Granados 2014; fig. 1.2). They've only continued since (Zettlemoyer et al. 2023).

As with the shift toward research on global change and other applied topics, the shift toward more-advanced statistical methods involves both diversification and replacement, but mostly the latter. It's not that ecologists these days now use hierarchical mixed models, Akaike information criterion (AIC), and Bayesian statistics in addition to ANOVA, it's that we use those more-advanced statistical approaches instead of ANOVA (Touchon and McCoy 2016; fig. 1.2).

The trend toward more-advanced statistics also shows up in the text-mining analysis of Anderson et al. (2021). Look at figure 1.1 again: two of the most rapidly rising 2-grams in the ecological literature of the 2000s are "model selection" and "explanatory variable." You don't have much use for either of those terms if you're only doing linear regressions and ANOVAs. A third statistical term, "meta-analysis," is another rapidly rising 2-gram (fig. 1.1). It deserves its own section, so we'll talk about it next.

Figure 1.2. Changes over time in the occurrence of search terms associated with six different statistical techniques, in seven leading ecology journals, 1990–2013

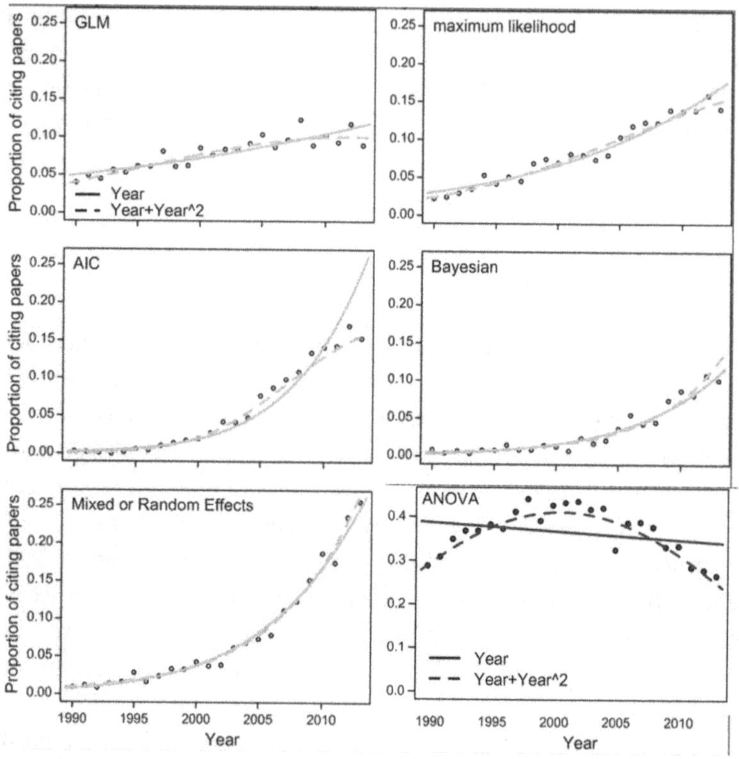

Source: Modified from Touchon and McCoy (2016).

Big Change 3: Meta-analysis

Meta-analysis is "an analysis of analyses." It's a family of statistical techniques for summarizing and comparing the results of different studies of the same topic. The underlying statistical ideas trace back at least to the early 20th century (O'Rourke 2006). Meta-analysis in its modern form developed in medicine in the 1970s as a way to summarize the results of different clinical trials of the same medication (O'Rourke 2006). The first ecological meta-analysis was published in 1991, but Jessica Gurevitch's work really put meta-analysis on ecologists' collective radar (Gurevitch et al. 1992; Hedges, Gurevitch, and Curtis 1999; Rosenberg, Adams, and Gurevitch 1997). Meta-analysis is now the standard way ecologists review and summarize the literature: in 2022 alone, ecologists published more than 160 meta-analyses, as indicated by a Web of Science search. Narrative reviews are now the exception rather than the rule—another example of diversification eventually turning into replacement. Indeed, in just 30 years, meta-analysis became so common that a few prominent ecologists began worrying that meta-analyses would replace not just narrative reviews, but papers reporting newly collected data (Lindenmayer and Likens 2011, 2013)! That worry is baseless, as we'll see later in this chapter; ecologists are collecting and publishing just as much data as ever. But that the worry exists at all is a striking (albeit anecdotal) illustration of just how important meta-analysis has become in our field.

Big Change 4: Data Sharing

Back when Gurevitch and others were pioneering the application of meta-analysis to ecology, they had to compile data by retyping printed data tables, digitizing figures, and writing to authors to request raw data files. That was tedious work, and often didn't work at all. In the early 1990s, it was common for ecology papers not to provide the data required for a meta-analysis, as well as for authors to ignore or decline requests to share their data files.

Times have changed. For instance, *every* empirical paper in the January 2024 issue of *Ecology* (the most recent issue as of this writing) provided its data in a public online repository for anyone to download and use.

That's an anecdote, but more systematic surveys show the same thing: ecologists have adopted online data sharing rapidly and widely. Michener (2015) is a good review of the early history of data sharing in ecology, but I'll highlight a few key events. The journal *Ecology* began publishing Data Papers in 2005. Data Papers are a way to publish datasets for other ecologists to reuse. They provide an incentive to authors to share their data. A Data Paper is a citable, peer-reviewed publication you can list on your CV along with your other publications. But many datasets of interest to other ecologists are associated with ordinary research papers rather than stand-alone Data Papers. So, in 2010 several leading ecology and evolution journals developed a Joint Data Archiving

Policy. In response to that policy, numerous ecology and evolution journals began mandating that authors share the raw data underpinning their papers, in a public repository accessible to anyone with an internet connection. As one would expect with any new policy, compliance was imperfect early on. Roche et al. (2022) reports that, as of 2012–2013, only 44% of ecology papers archived complete data, and only 36% archived their data in such a way as to present no obstacles to reuse. But given that less than 10 years earlier, both of those numbers would've been approximately 0%, the word "only" arguably doesn't belong in that previous sentence.

Routine data sharing is another example of replacement rather than diversification in ecological research. Shared data has replaced unshared data.

Rapid adoption of meta-analysis and data sharing went hand in hand. Indeed, they likely were mutually reinforcing. More and more ecologists wanted to read and conduct meta-analyses, which created demand for data sharing so that meta-analyses would be easier to conduct. Conversely, easier availability of more data likely encouraged more ecologists to conduct more meta-analyses than they otherwise would have.

Big Change 5: Collaboration

When I started graduate school in 1995, it wasn't all that unusual for ecologists to do as I did early in my career and work solo. Half of the twelve papers I published from work I did as an undergraduate, graduate student, or postdoc were sole-authored. At the time, that made me a bit old-fashioned, but not an outlier. For instance, approximately 10–15% of papers published in *Ecology* and *Journal of Ecology* in the 2000s were sole-authored (Gorham and Kelly 2014). Back then, a plurality of papers were written by either a single author or a pair of authors (often a current or former trainee, together with their current or former adviser). Only a few papers (<10%) had more than five authors, indicative in many cases of large collaborations involving authors at multiple institutions (Gorham and Kelly 2014).

That was then, this is now. Collaborative ecological research (a category that includes collaborative meta-analyses) is now the norm. I checked the January 2022 issue of *Ecology* and found that the average paper had 5.4 authors from 2.4 institutions. Big systematic studies tell the same story: across many fields of science, social science, and humanities, including ecology, multiauthored papers have been increasing in frequency since at least 1900, and have outnumbered sole-authored papers since 1980 (J. D. West et al. 2013; Gorham and Kelly 2014; Duffy 2017; Zettlemoyer et al. 2023). Ecological papers with more than 20 authors—all but unheard of when I started graduate school—are now a thing (C. Fox, Paine, and Sauterey 2016). This shift partially reflects loosening standards for authorship (e.g., Duffy 2015), but it also reflects a real increase in collaborative work.

Big Change 6: Geographic Diversification

One axis along which ecological research has diversified in a big way during my career is geographic diversity. Plenty of barriers to geographic diversification remain (Nuñez et al. 2019; Pettorelli et al. 2021). But one can acknowledge those barriers, and work to dismantle them, while also recognizing how many barriers have already come down. As with data sharing, this is another change that's so big you don't have to compile much data to detect it. For instance, I checked every paper published in the first issue of *Ecology* for the years 1982, 2002, and 2022. For every paper, I checked whether the paper reported data collected by the authors as opposed to compiled from the literature. If the paper reported data collected by the authors, I recorded which country/countries and continent(s) the data came from. All three issues had similar numbers of suitable papers, so any temporal trends in geographic diversity of data sources wouldn't be confounded with the number of published papers. Fully 96% of papers in the February 1982 issue of *Ecology* reported data from North America, mostly the US. Only one paper out of 26 reported data from elsewhere (a single country in Latin America). Fast-forward to the January 2002 issue of *Ecology*: only 70% of papers reported data from North America, with 40% reporting data from elsewhere (a few reported data from both within and outside North America). Cumulatively, the 18 suitable papers in that issue reported data from six countries on five continents, up from just three countries on two continents back in 1982. Fast-forward to the January 2022 issue of *Ecology*: only 42% of papers reported data from North America. They were outnumbered by the

63% that reported data from outside North America. This wasn't just a matter of *Ecology* adding a bunch of papers with data from Europe, either: 50% of papers in the January 2022 issue of *Ecology* reported data from continents other than North America or Europe. Cumulatively, the 24 suitable papers in that issue reported data from 22 countries on six continents. There were similar trends in *Ecological Applications* from January 2002 to January 2022: data from four countries on two continents in 2002, increasing to data from 12 countries on six continents in 2022.

Field studies in ecology still don't consider a random sample of the world's locales, of course. Not that we'd want them to, obviously—at least, not for most purposes. The goal of overcoming undesirable constraints and biases in ecologists' choices of study taxa and study sites is not to randomly or uniformly sample all taxa or sites. Rather, the goal is to maximize ecologists' ability to focus our collective research efforts in desirable ways. Terrestrial studies tend to focus disproportionately on sites located in protected areas, in forests, in temperate climates, in wealthy countries, close to universities (Pyšek et al. 2008; Pokallus et al. 2011; Martin, Blossey, and Ellis 2012; Trimble and van Aarde 2012; Archer et al. 2014; Meyer, Weigelt, and Kreft 2016; Di Marco et al. 2017; Crystal-Ornelas and Lockwood 2020a; White et al. 2021). Those terrestrial studies that do come from the tropics tend to come disproportionately from protected sites in comparatively wealthy tropical countries, especially sites located close to field stations or roads (Reddy and Dávalos 2003; Stocks et al. 2008; Powers et al. 2011; Pitman et al. 2011; Cosentino and Maiorano 2021; Gross and Heinsohn 2023). Marine studies are concentrated in less human-impacted areas of the nontropical Northern Hemisphere (e.g., Mott and Clarke 2018). However, the few papers that look for temporal trends in geographic biases find that those biases have decreased over time, with historically understudied locations receiving growing research attention relative to historically well-studied locations (e.g., Di Marco et al. 2017).

Geographic diversification of ecological research reflects geographic diversification of the ecologists who do that research. As a small illustration, consider again the first issue of *Ecology* from the years 1982, 2002, and 2022. Every author of every paper in the February 1982 issue of *Ecology* was based in the US. In contrast, the January 2002 issue of *Ecology* included papers by authors based in seven countries. In greater contrast, the January 2022 issue of *Ecology* included papers by authors based in 16 countries. That drove geographic diversification of data collection: for 85% of papers across those three issues of *Ecology*, there was overlap between countries in which authors were based and countries from which the data were collected. The geographic diversification of authors in turn reflects the increased collaborativeness of ecological research, noted above. The more authors there are per paper, the more countries you'd expect those authors to come from, all else being equal. OK, those data only concern one journal. But I doubt geographic diversification is unique to the journal *Ecology*, because larger datasets show the same thing. For instance, Brandell et al. (2021) reports survey data indicating increasing geographic diversification of self-identified disease ecologists. Pokallus et al. (2011) describes geographic diversification and collaboration among US ecologists.

In summary: ecologists now collaborate more, to study more applied topics, in more places, using more-advanced statistics. Those are big changes! And they've happened within the professional lifetimes of many ecologists working today. Without wanting to downplay the magnitude or importance of those changes at all, let's now turn our attention to what *hasn't* changed.

BIG STASES

I'd say what hasn't changed are ecologists' research goals and research approaches (the ways we go about achieving our research goals). The quickest way to clarify what I mean by these terms is to provide a coarse classification of each.

I divide "research goals" into four broad categories. Here they are, in no particular order:

Description. The goal of the research is to describe how the world is. How many species are there on the planet (Wiens 2023)? What is the global distribution of plants used by humans (Pironon et al. 2024)? How are species' geographic ranges shifting in response to climate change (Rubenstein et al. 2023)? Is the average effect of experimental mycorrhizal fungal inoculation on plant growth positive or negative (Hoeksema et al. 2010)?

Explanation or understanding. The goal of the research is to answer a "why" question or to solve a puzzle. Why do some populations exhibit cycles in abundance (Barraquand et al. 2017)? Why is resting metabolic rate proportional to body mass raised to the 3/4 power (G. West, Brown, and Enquist 1997)? How do competing species manage to coexist for many generations, without one species eventually excluding all the others (Chesson 2000)? Often, answering a "why" question or solving a puzzle requires figuring out causality. Sometimes the answer involves showing that the "why" question was ill posed, meaning it doesn't have an answer (Loehle 2011). Sometimes, the answer involves showing that the purported puzzle isn't actually a puzzle at all. Rather, it only appears puzzling because we're making unwarranted assumptions or failing to consider relevant information.

Prediction or forecasting. The goal of the research is to predict as-yet-unobserved data. Is the orchid *Angraecum sesquipedale* pollinated by an as-yet-undiscovered long-tongued moth (a famous prediction of Charles Darwin's; see Arditti et al. 2012)? How will species' geographic ranges shift in response to climate change (Morin and Thuiller 2009)? Will North Atlantic cod stocks ever recover from overfishing, and if so, when (Swain and Chouinard 2008)?

Manipulation, control, restoration, or management. The goal of the research is to enable someone to change the world in some way or to prevent it from changing in some way. Where should additional dollars of conservation spending be directed so as to maximize impact per additional dollar spent (Waldron et al. 2017)? What goals do ecological restoration plans typically have, how is restoration success evaluated, and how common is success (Wortley, Hero, and Howes 2013)? Would livestock culling prevent the spread of foot-and-mouth disease (Haydon, Kao, and Kitching 2004)? Is the tope shark "endangered" in the US, as that term is defined by the US Endangered Species Act (NOAA Fisheries 2023)?

These goals aren't mutually exclusive. Any given ecologist, or any ecological research paper, can have multiple goals. And there are contexts in which these goals are complementary. For instance, it's pretty difficult to explain, forecast, or manage a phenomenon you can't even describe accurately. As another example, predictions can be used both to test whether we've correctly explained some phenomenon, and to inform management decisions by predicting what the consequences of those decisions will be. Nor are any of these goals "better" or "more important" than any of the others; there's no context-independent, universal ranking of these goals. There might be trade-offs between achieving different goals. Finally, different goals might be complementary, so that the "best" research approach will combine different goals. Different ecologists naturally will differ in the goals they want to achieve, or are able to achieve, for all sorts of reasons—different preferences, different skills, and so on. Not every ecological researcher is always going to be pursuing every goal on my little list. That makes it interesting to ask whether some research goals are more popular than others, and how the popularity of different goals has (or hasn't) changed over time.

Ecologists use various research approaches to achieve their goals. I divide "research approaches" into eight broad categories. Here they are, in no particular order:

Collecting observational field data. That is, you collect data out in the field, without intervening to change anything. Whatever variation there is among your observations, it exists independent of your efforts to collect those observations. Note that, depending on the goal of the study, the "field" might not be a "natural" or even seminatural setting. You can collect ecological "field" data from, say, people's gardens, or rivers in urban areas, or on the microbes living in water treatment facilities, and so on. Note as well that collecting observational field data sometimes involves bringing samples back to a laboratory. For instance, you might clip grasses at ground level, then dry the samples in an oven and weigh them in the lab to record their dry biomass. The dry biomass of grass can't be measured in the field, so it has to be measured in the lab. Finally, note that the observer doesn't have to be physically present to collect observational field data. You can collect observational field data with satellites, drones, or camera traps. Or you can recruit others to collect observational data for you, as in many citizen science projects.

Conducting a field experiment. That is, you intervene to change something out in the field. The intervention creates some observable variation in nature that wouldn't have existed otherwise.

Conducting a laboratory/greenhouse/microcosm/mesocosm experiment. Rather than conducting an experiment in the field, you conduct your experiment in some artificial setting. This allows you to eliminate or regulate many sources of variation that would ordinarily be present in any field setting. The artificial setting may also allow you to manipulate or measure some variable that would be infeasible or unethical to manipulate or measure in the field.

Conducting some other kind of study, besides field observations, field experiments, or laboratory/greenhouse/microcosm/mesocosm experiments—for instance, by compiling data from museum specimens.

Compiling and synthesizing data from the scientific literature, and/or other sources, rather than collecting data yourself.

Building and analyzing a mathematical model. Rather than study actually existing examples of whatever it is you're studying, study a mathematical version. Note that this can involve expressing the model in the form of equations and other mathematical objects, which you study using mathematical tools such as algebra and calculus. It can also involve expressing the model in the form of a computer program, which you study by running the program and examining its output.

Proposing a new method for performing some scientific task. There is an entire journal, *Methods in Ecology and Evolution,* devoted to papers that do this.

Sharing an opinion. Try to persuade other ecologists to change their research however you think it should change.

It's more difficult to document changes (or the lack thereof) in ecologists' research goals and approaches than it is to document changes in paper topics or statistical methods. That's because research goals and approaches are harder to study via text mining and other automated methods. So below I mostly rely on hand-compiled data, from myself and others.

It's worth noting that my fairly coarse taxonomy of research goals and approaches won't reveal changes within each taxonomic category. For instance, a small but rapidly increasing fraction of descriptive studies in ecology now use citizen science: nonprofessional scientists collecting data in collaboration with, or under the supervision of, professionals (Pocock et al. 2017). "Fine-grained" changes in our research goals and approaches, such as the rise of citizen science, happen all the time. Those sorts of changes are important, but they're also the sorts of changes that professional ecologists tend to be well aware of as they're happening. In contrast, I think that we're collectively less alert to coarser-grained changes, or the lack thereof. Indeed, precisely because we're all well aware of the constant flood of new "fine-grained" ideas and approaches reported in the research papers published every day, we may not realize just how static ecological research is at a coarse-grained level.

My own data compilation was admittedly haphazard, but not so haphazard as to make it useless, I don't think. I looked at all the papers in the first issue of the year for each of several ecology journals, for the years 2002 and 2022 (and 1982, for journals that went that far back). I picked the journals *Ecology*, *Oikos*, *Ecological Applications*, *Ecology Letters*, and *Biotropica* so as to have a mix of journals that varied on various dimensions (e.g., applied vs. not; range of impact factors; geographically focused vs. not, etc.). I recorded various bits of information about those papers.

Big Stasis 1: Lack of Hypotheses

Here's one non-trend backed by a lot of data: ecologists rarely test scientific hypotheses (as distinct from statistical null hypotheses). This implies that ecologists rarely conduct research with the goal of explanation, since scientific hypotheses can be defined as proposed explanations (Betts et al. 2021). A statistical null hypothesis like "The true population correlation between variables *x* and *y* is zero" isn't a proposed explanation, though it could be one line of evidence for a proposed explanation.

It might seem surprising that ecologists rarely test scientific hypotheses. In my experience, ecology graduate students have the importance of hypothesis testing drummed into them. I doubt my experience is all that atypical. You might think that if we're all taught to test scientific hypotheses, we actually do so. But we don't. Sometimes the world is surprising!

The authors of Betts et al. (2021) closely examined a stratified random sample of 268 ecology and evolution papers from 22 journals, published from 1990 to 2015. They also searched for occurrences of "hypoth*" in the titles or abstracts of many thousands of papers from those same journals. Only about 25% of papers in the closely examined sample had any scientific hypotheses at all, and the frequency of papers with hypotheses didn't change significantly over time (fig. 1.3). Papers comparing multiple alternative hypotheses were especially rare. The same results held in the larger sample. Fewer than 10% of ecology papers published between 1990 and 2015 had "hypoth*" in the title or abstract, and their frequency hardly budged over time. The fact that the same results held in both samples indicates that it's not the case that many ecology papers test scientific hypotheses without using the word "hypothesis" in the title or abstract. Finally, a substantial minority of hypotheses in ecology papers are what the authors termed "descriptive hypotheses." A descriptive hypothesis isn't really a scientific hypothesis. Rather, it's just a descriptive statement of one of the possible outcomes of the study, with the word "hypothesis" tacked on. A descriptive hypothesis lacks any explanation as to why the hypothesized outcome might be expected, or why one possible outcome is arbitrarily singled out as the "hypothesized" outcome.

Even if you restrict attention to leading journals, many ecology papers lack any scientific hypotheses (again, not counting statistical null hypotheses), and few compare multiple scientific hypotheses. Gustavo Betini, Tal Avgar, and John Fryxell (2017) examined 20 randomly selected papers from each of five leading ecology and evolution journals, published from 2000 to 2011. Of those 100 papers, 41 either had no hypotheses at all, or the statistical null hypothesis was the only hypothesis. Of the remaining 59 papers, only 21 considered multiple scientific hypotheses. Applied papers were particularly likely to be hypothesis-free. Given the increasing frequency of applied papers in the ecological literature, one wonders if hypothesis testing will become even rarer than it already is.

Other studies find similar results. Paul Grogan (2005) sampled 98 papers published in 2004 in various leading ecology journals and found

Figure 1.3. Ecologists rarely test alternative scientific hypotheses

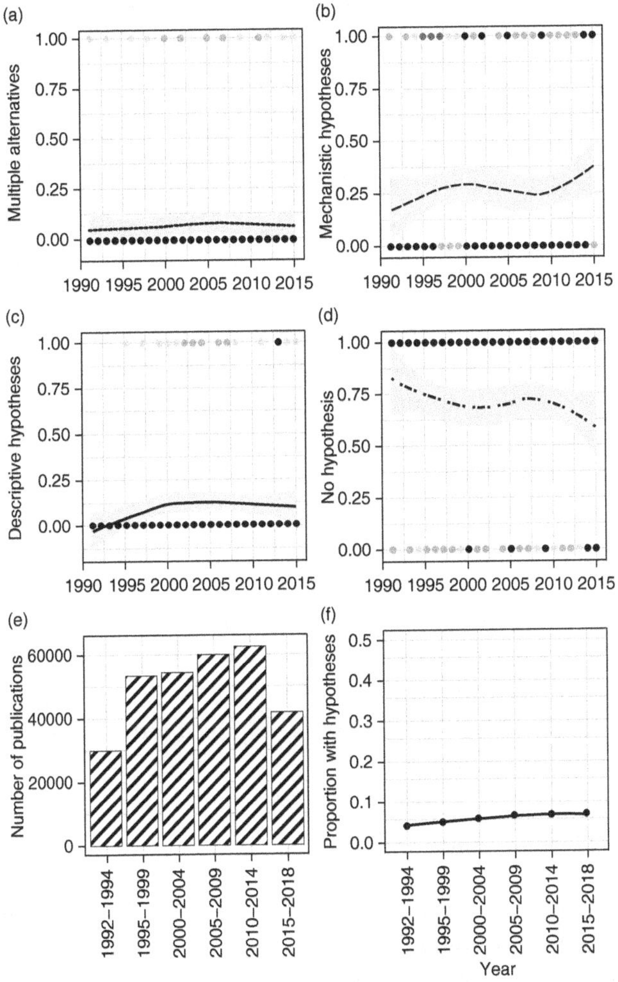

Trends in hypothesis use from 1991 to 2015 from a sample of the ecological and evolutionary literature ($N = 268$). Panels a–d show estimated frequencies over time of papers with (a) multiple alternative hypotheses, (b) mechanistic hypotheses, (c) descriptive hypotheses, and (d) no hypotheses. Lines reflect LOESS smoothing with 95% confidence intervals and indicate no significant temporal trends. Dots show raw data, with darker shades indicating overlapping data points. The total number of publications in ecology and evolution in selected journals has increased (e), but use of the term "hypoth*" in the title or abstracts of these publications is very low and has hardly increased over time (f).

Source: Figure and caption modified from Betts et al. (2021).

that less than half of their Introduction sections stated any hypotheses at all (not even statistical null hypotheses). Charles Krebs (2015a) examined every paper from the first two issues of *Journal of Animal Ecology* published in 2015, asking if the paper explicitly stated alternative hypotheses. Only 11 out of 51 papers (22%) did so. Erlend Nilsen, Diana Bowler, and John Linnell (2020) examined 159 terrestrial ecology papers randomly sampled from eight journals covering applied ecology, wildlife management, and conservation biology. Only 23% explicitly stated any hypotheses at all (an additional 28% implied a hypothesis). Only two of the 159 papers explicitly stated alternative hypotheses. The authors of Sells et al. (2018) reported that only 44% of papers published in the *Journal of Wildlife Management* from 2013 to 2016 stated or implied any hypotheses at all, never mind alternative hypotheses.

But wait: what about the rise of AIC and other statistical methods of model selection (see Big Change 2 and fig. 1.2, above)? Isn't statistical model selection a way of comparing alternative scientific hypotheses? Yes and no. It certainly can be—see, for instance, the examples from behavioral ecology compiled in Dochtermann and Jenkins (2011). But more often, ecologists use AIC and other model selection methods merely as a way of identifying predictor variables correlated with variation in the dependent variable—often with no scientific hypotheses, or only vague scientific hypotheses, as to which correlations to expect or why (McGill 2015).

Big Stasis 2: Many Observational Field Studies

Back in, say, the 1950s, most ecology papers were observational field studies. Paper titles of the form "The Ecology of [name of species]" were common; the paper would report field observations about [name of species]. Our paper titles have changed. These days, even the rare papers on "The Ecology of [name of species]" aren't titled that way. But collecting observational field data is still the most common research approach in ecology (fig. 1.4). And it hasn't become any less common over the last few decades (fig. 1.4). The same is true if you restrict attention to the applied research topics that increasingly dominate the literature: observational studies far outnumber experimental studies in applied ecology journals (Nilsen, Bowler, and Linnell 2020).

Figure 1.4. Predominance of field observations as a research approach in ecology

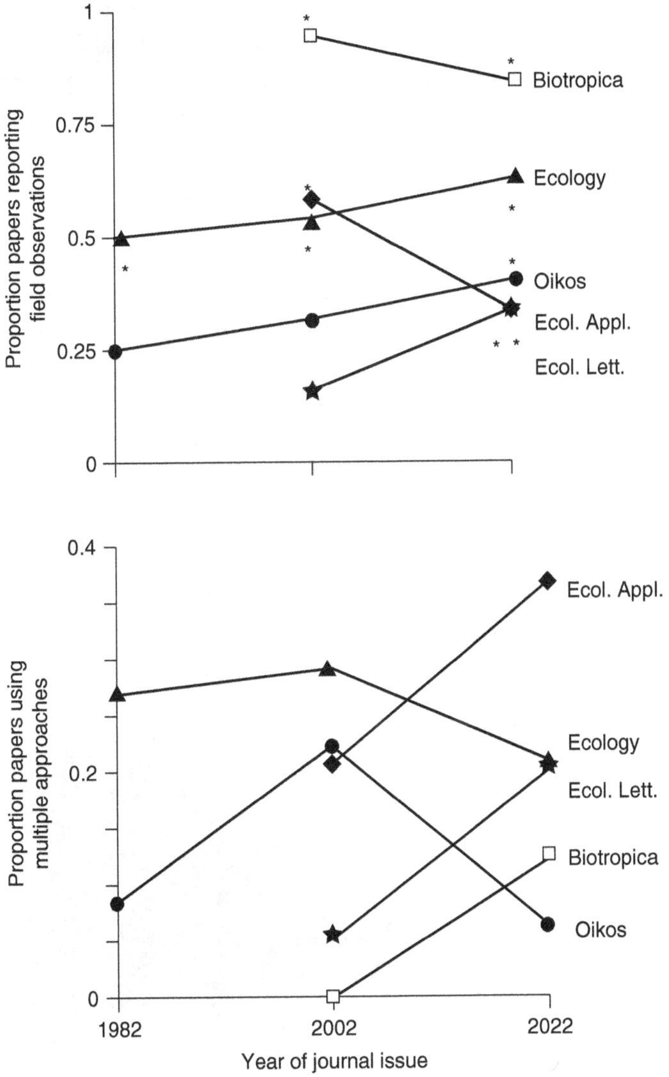

Top: Proportion of papers in the first issue of the year that reported observational field data collected by the authors, for various ecology journals. Asterisks mark those journal issues for which collecting observational field data was the most common research approach. *Bottom*: Proportion of papers in the first issue of the year that used multiple research approaches, for various ecology journals.

The predominance of observational field studies as a research approach doesn't necessarily imply anything about ecologists' goals. One could (and ecologists sometimes do) collect observational field data in order to make a prediction, test a hypothesis, inform a management decision, and so on. But in my experience, most ecology papers reporting observational field data have description as their primary goal.

Big Stasis 3: One Research Approach per Paper

Of course, a paper that reports observational field data might report other sorts of data too. More broadly, any ecology paper can report results from multiple research approaches. But they mostly don't, and they mostly never have (fig. 1.5). A substantial majority report results from only a single research approach. That's even though many ecology

Figure 1.5. The typical sample size in ecology has not increased over time, and may have declined

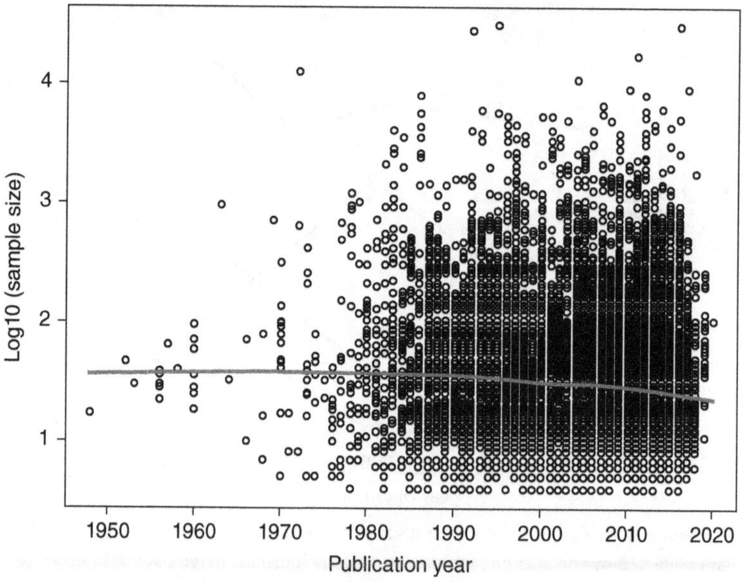

Log_{10}-transformed samples sizes of correlation coefficients that were later included in ecological meta-analyses, versus their year of publication. Many points are identical and so not visible. The gray line is a LOESS smooth.

Source: Data from Costello and Fox (2022).

papers these days relegate many methodological details and results to lengthy supplements. Thanks to supplements, many ecological papers contain much more text and many more figures than they used to. But that extra text and those extra figures don't translate into extra research approaches. Rather, they translate into more-detailed reporting of the results from a single research approach.

Big Stasis 4: No Mathematical Models outside of Modeling Papers

Some research approaches, such as collecting observational field data, have remained consistently common in ecology for decades. Other approaches have remained consistently rare. Ecological papers rarely include mathematical models other than statistical models—unless they're mathematical modeling papers, of course. I examined a total of 256 papers from an admittedly small, haphazard set of empirically focused ecology journals (the first issues of *Ecology*, *Ecological Applications*, *Oikos*, *Ecology Letters*, and *Biotropica* from 2002 and 2022, plus the first issues of *Ecology* and *Oikos* from 1982). Only 27 out of 256 (11%) included mathematical modeling among their approaches. Only 16 out of 256 (6%) included mathematical modeling plus at least one other approach. And there was no hint of any upward trend over time in the prevalence of mathematical modeling.

Of course, perhaps that's because those journals are all widely seen as empirical journals rather than theory journals. But that's precisely the point here. Mathematical modeling in ecology is largely restricted to theory journals. With the arguable exception of *American Naturalist*, ecology journals are all seen as *either* empirical journals *or* theory journals. For the most part, ecologists don't write papers that combine mathematical modeling with other research approaches, and the rare papers that do so don't have many natural homes among ecology journals.

Big Stasis 5: Small Sample Sizes and Low Statistical Power

OK, so ecologists are still conducting lots of observational field studies, just as we always have. And we're still conducting smaller numbers of field experiments, laboratory experiments, and the like. But surely those studies themselves are improving? In particular, surely their sample sizes and statistical power are increasing?

After all, we now have various technologies that we didn't have decades ago, that allow us to collect, or compile, many more observations of many more variables at many more locations than we could before. We have satellite remote sensing of landscape cover, camera-trap networks for detecting and counting animals, GPS collars for tracking animal movement, digitized museum specimens, online compilations of phenotypic trait measurements for thousands of species, cheap whole-genome sequencing, and on and on (Kattge et al. 2011; Pettorelli et al. 2021; Steenweg et al. 2017; Nelson and Ellis 2018; Costa-Pereira et al. 2022; N. Harrison and Kelly 2022; Castagneyrol, Bedessem, and Julliard 2023). In inflation-adjusted terms, we also have much more government and nongovernmental funding for ecological research than we had back in, say, the decades immediately after World War II (NSF, n.d.-a; AAAS, n.d.). And it's now routine for ecologists to take courses in statistics, which wasn't the case in the decades immediately after World War II. Surely ecologists these days are all well aware of the importance of sufficient sample size and statistical power, and design their studies accordingly?

Well, awareness might be higher than it was several decades ago—but our actual sample sizes and statistical power aren't. Empirical studies in ecology mostly have small sample sizes. Those sample sizes haven't increased since before I was born, never mind since I became an ecologist. Sample sizes are especially small relative to the magnitude of effect ecologists typically study. This means that ecological studies typically have low statistical power. Further, even when a low-powered study does detect a statistically significant effect, the absolute magnitude of the effect is likely to be greatly overestimated.

Let's unpack that last paragraph. Ecological studies typically report small sample sizes—often, almost as small as they possibly could be without giving up on replication entirely! Experiments on multiple stressors in aquatic ecosystems have a median sample size of $n = 3$ replicates per treatment, and only 1% of experiments have $n > 6$ replicates per treatment (Burgess, Jackson, and Murrell 2022). Experiments in landscape ecology aren't much better: their median sample size is $n = 6$ replicates per treatment (Wiersma 2022).

It's not just experiments that use small sample sizes. Laura Costello and I (Costello and Fox 2022) attempted to compile a comprehensive database of all empirical ecological studies, on any ecological topic, that were later included in meta-analyses. Our compilation included more than 16,000 correlation coefficients, from more than 5,000 primary research papers included in 232 meta-analyses on a wide range of ecological topics. Most correlation coefficients in ecology come from observational studies. The modal sample size is $n = 10$ observations, and the median is $n = 30$.

Nor have sample sizes increased over time in ecology. Figure 1.5 plots the sample sizes of those 16,000-plus correlation coefficients against their year of publication. The nonparametric regression line suggests that if anything, the typical sample size in ecology has *declined* over time. Don't take that regression line too seriously; it doesn't allow for non-independence. It's just there to confirm what your eyes (or a more rigorous statistical analysis) tell you: for the most part, ecological sample sizes are small and haven't increased since records began.

Those small sample sizes don't provide much statistical power. Sticking with the correlation coefficients from figure 1.5, the modal sample size of $n = 10$ provides ≥80% power against a null hypothesis of zero only if the absolute value of the true correlation is greater than ~0.8 (Bujang and Baharum 2016). The median sample size of $n = 30$ confers ≥80% power against a null hypothesis of zero only for true correlations greater than ~0.5 or less than ~−0.5 (Bujang and Baharum 2016). In practice, most correlations that ecologists study surely are smaller than 0.5 in absolute magnitude, so statistical power is ordinarily much less than the widely recommended value of 80%.

Ecological experiments also have low power, and there's no sign that power has increased in recent decades. The authors of N. Lemoine et al. (2016) and Yang, Hillebrand, et al. (2022) calculated the power of field experiments on global change–related variables (drought, warming, acidification, fire, biodiversity loss, others) to detect typical effects of those variables. Both studies defined a "typical" effect as the statistically average effect, as estimated from meta-analyses of the effects of those variables. Power to detect a typical effect, against a null hypothesis of zero effect, ranged from 6–38% depending on the variable considered. Further back, Michael Jennions and Anders Møller (2003) found that statistical tests in behavioral ecology averaged only 13–16% power to detect a "small" effect (e.g., a correlation of 0.1), and 40–47% power to detect a "medium" effect (e.g., a correlation of 0.3). Similarly low power shows up in everything from animal tracking studies (Cleasby et al. 2021) to studies of the effects of toe clipping on anuran survival (Parris and McCarthy 2001).

Low power also has a pernicious side effect: it generates "Type M errors" (Gelman and Carlin 2014). Type M errors are "magnitude" errors—overestimation of the absolute magnitude of a population parameter. Imagine that, against the odds, a low-powered study does produce a statistically significant effect. Assume further that this isn't a Type I error; the null hypothesis of zero effect actually is false. The odds are that the estimated effect size will be much larger in absolute magnitude than it truly is, because if the effect size hadn't been overestimated, it likely wouldn't have come out statistically significant. That is, the statistically significant estimates are an upwardly biased subset of all the estimates that the study would produce if it were repeated many times. Lemoine et al. (2016) report that, if a typical field experiment on a global change–related variable is to come out statistically significant, it likely has to overestimate the absolute magnitude of the true effect by a factor of 1.5–3—a substantial Type M error.

EXTERNAL VERSUS INTERNAL DRIVERS OF CHANGE AND STASIS

Look back over those lists of big changes in ecological research over the past few decades. What do they have in common? I think the common thread is that the big changes were all driven by, or at least enabled by, factors external to the field of ecology.

The rise in research on global change and other applied areas started in the late 1980s and early '90s. It followed on the heels of, or coincided with, a massive global rise in public and political interest in climate change, particularly global warming. The Intergovernmental Panel on Climate Change (IPCC) was established by United Nations (UN) resolution in December 1988 and issued its first Assessment Report in 1990. *Time* magazine—then a popular and influential weekly news magazine in the US—named "Our Endangered Planet" its "Person of the Year" for 1988. The 20th Earth Day, in 1990, was front-page news around the world and paved the way for the UN Earth Summit in 1992. It would've been surprising if the topics of ecological research *hadn't* shifted in response.

The massive shift toward more-advanced statistical approaches that took off in the early to mid-2000s was enabled in large part by the advent of R: a powerful, free, open-source software package (Touchon and McCoy 2016; Lai et al. 2019; fig. 1.6). This isn't to say that internal discussions among ecologists had no role. In particular, the increased popularity of AIC within ecology may have been due in part to the influence of Burnham and Anderson (2003). But many fields of science and social science have shifted toward more-advanced statistical approaches, alongside widespread adoption of R (Touchon and McCoy 2016; Lai et al. 2019; Healy 2018; M. Clark, n.d.). So the shift toward more-advanced statistical approaches likely wasn't due primarily to factors internal to ecology.

The increasing geographic diversity of ecological research and researchers—more authors from outside the US and Europe, publishing more papers that report data from outside the US and Europe—also

Figure 1.6. R is now the dominant statistical software in ecology

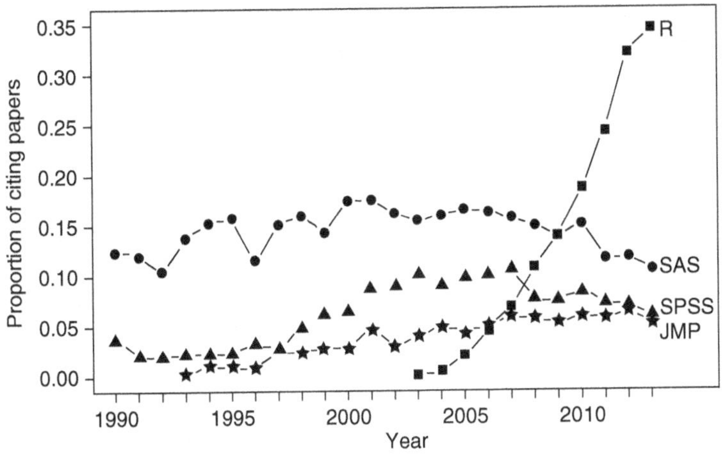

Changes over time in the frequency of citations of four statistical software packages (R, SAS, SPSS, JMP) in seven leading ecology journals, 1990–2013.

Source: Modified from Touchon and McCoy (2016).

seems driven primarily by factors external to ecology. Scientific research costs money and other resources, and requires people with the specialized skills and knowledge to do it. Many countries outside the US and Europe are now much more populous, much wealthier, and much better educated than they were in, say, 1990 (Ritchie et al. 2023; "GDP per Capita," n.d.; "Share of the Population with Tertiary Education," n.d.). So they're doing, and publishing, a lot more scientific research than they used to in all fields, not just in ecology (NSF, n.d.-b).

The increasing trend toward collaboration among ecological researchers also seems like a response to external factors. Researchers in all scholarly fields are becoming more collaborative, including across institutions, in large part because the internet eases collaboration. Email first came into widespread use among researchers in all fields in the 1990s. This was followed by a steady stream of other technological advances that make remote collaboration easier (faster internet connections, file sharing, cheap webcams, free videoconferencing software, cloud storage . . .).

The one partial exception might be the rapid uptake of data sharing and meta-analysis. I'm sure this has been driven in part by external factors. The internet eases the data sharing on which meta-analysis depends. And the advent of powerful, widely available statistical software packages for meta-analysis (R, and before that MetaWin; Senior et al. 2016; Rosenberg, Adams, and Gurevitch 1997) made the growth in ecological meta-analysis possible. But not every field of science has taken up data sharing and meta-analysis in a big way, and those that did take them up didn't all do so simultaneously. The internet and new statistical software were likely necessary but not sufficient conditions for the rapid uptake of data sharing and meta-analysis in ecology. That suggests an important role for factors internal to our field in explaining why ecologists embraced data sharing and meta-analysis so quickly. I speculate that one internal factor is the influence of the National Center for Ecological Analysis and Synthesis (NCEAS), founded in 1995. NCEAS was founded on the belief that ecologists could learn a lot from data we'd already collected, if only different datasets could be compiled in a useful form (e.g., in a single electronic database, rather than scattered across printed tables in separate scientific papers).

Now look back over the list of things that haven't changed. What do they have in common that distinguishes them from the things that have changed? I'd say it's the lack of any external forces driving or enabling change. There haven't been any big external changes in technology, economics, or society that would prompt ecologists to, say, diversify away from reliance on observational data. And there's no reason why any of the big external drivers of change in some aspects of ecological research should also have changed other aspects. Nothing about global economic growth encourages ecologists to adopt explanation as a research goal. The internet doesn't create any strong incentives for ecologists to test multiple alternative hypotheses. There's no reason why worldwide concern about climate change would enable empirical ecologists to make heavier use of mathematical modeling. Widespread adoption of R doesn't facilitate combining multiple research approaches. And so on.

Which raises the question of why I bothered to write this book! Why add to the pile of doomed attempts to drive fieldwide change in ecology internally? Because my hopes and ambitions for this book are more modest than that. I do think the field of ecology needs to change. But I am under no illusions that this book—or any book—will reach enough readers, or have a big enough influence on the readers it does reach, to have any detectable effect on the direction of the entire field. Nobody is going to update text-mining studies like Anderson et al. (2021) in 25 years and discover a signal arising from the publication of this book! The size and diversity of our field make it impossible for any one person to have any detectable influence on the field as a whole. But if this book reaches some modest number of readers who find it useful, thought-provoking, or even (dare to dream) inspiring, it will have achieved as much as any one book can reasonably be hoped to achieve in ecology.

Agreeing to Disagree?

So that's how the field of ecology has, and hasn't, changed in my professional lifetime. What do ecologists think about those changes? Are we all happy with the state of the field? Have we all now come to a consensus as to what ecology is and how to do it, setting aside the narrow technical disagreements that are a normal part of science?

Maybe! But I'm not so sure we've come to a consensus so much as we've all agreed to disagree. In 2019, I polled readers of the *Dynamic Ecology* blog as to their views on a list of 24 purported problems with ecological research (J. Fox 2019b). Are ecologists failing to develop and test hypotheses? Undervaluing system-specific case studies? Overvaluing meta-analysis? Overvaluing applied research? Undervaluing natural history? Or what? For each purported problem, respondents were asked if they considered it a serious problem, a moderate problem, not a problem, or the opposite of a problem. The 115 poll respondents obviously weren't a random sample—they were a self-selected subset of readers of one blog. But *Dynamic Ecology* has many readers from around the world (J. Fox 2020b, 2020a), and its poll respondents at that time were diverse on various dimensions. For instance, *Dynamic Ecology* readers at that time comprised a mix of grad students, postdocs, faculty, and nonacademic professional ecologists, and only about half were based in the US. So although this poll wasn't a random sample of ecologists, it was a large enough and diverse enough sample to be worth talking about.

Turns out that ecologists all agree that ecology has some serious problems, but everyone disagrees as to what they are. The median respondent identified four problems from the list of 24 as serious (mean 3.9, max 17). Only 19 out of 115 respondents didn't identify any of the problems on the list as serious. Every problem on the list was considered serious by at least 5% of respondents. But no problem was considered serious by a majority of respondents. If you convert the seriousness scale to integers, from 2 for "serious problem" down to −1 for "opposite of a problem," and plot the mean and variance of responses for each problem, you get a crude but useful graph (fig. 1.7).

Figure 1.7 is the graph you get if everyone has a different bee in their bonnet regarding the field of ecology.

But the variation in figure 1.7 isn't random variation. If you know the sort of research that ecologists do, you can predict their opinions as to the field's serious problems (J. Fox 2019b). For instance, the poll asked respondents if they do fundamental research, applied research, or a mix. Applied researchers are much more likely than fundamental researchers to see "valuing theory over data," "overvaluing novelty," and "valuing generality over case studies" as serious problems in ecology. To a first approximation, everyone thinks their own goals and approaches

Figure 1.7. Ecologists disagree as to the seriousness of most purported problems with ecological research

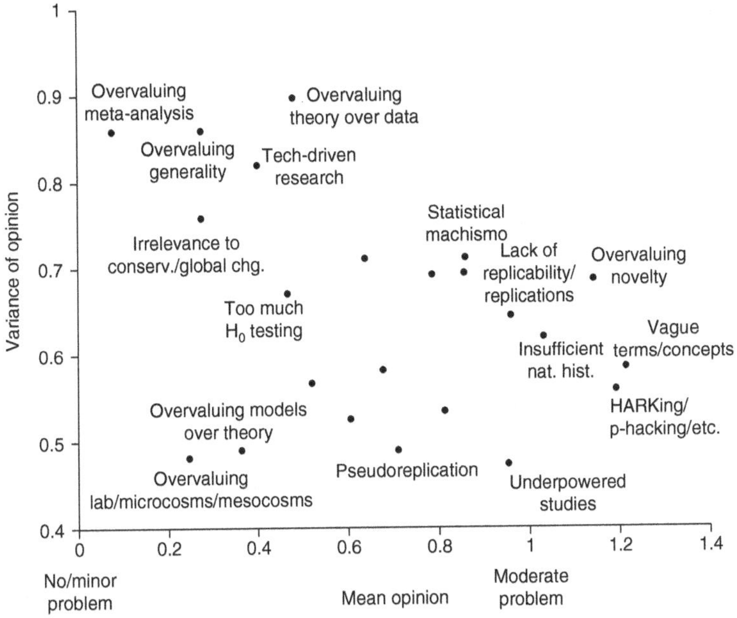

Each point gives the mean opinion, and variance of opinion, of 115 readers of the *Dynamic Ecology* blog regarding the seriousness of a purported problem with ecological research (J. Fox 2019b). Opinions about each problem were scored on a four-point scale (2 = serious problem, 1 = moderate problem, 0 = no/minor problem, −1 = opposite of a problem). Some of the problems are labeled.

are undervalued and that other people's goals and approaches are overvalued. So in that sense, there actually is some agreement here: many ecologists agree that their own research is great, and it's everybody else's research that's the problem.

So again, we ecologists haven't resolved our disagreements so much as we've agreed to disagree. Is *that* a problem? Well, there's something to be said for everyone just doing their own thing and letting everyone else do the same. It's probably pointless, and definitely no fun, to keep having the same arguments over and over about what ecology is and how to do it. But agreeing to disagree has its own costs. One is lost opportunities to learn from one another. Another cost is loss of a shared sense of participating in a common enterprise called "ecology"—a sense of ecology as a unified field of inquiry that's greater than the sum of its parts because those parts are interconnected in coherent ways. If everyone just does their own thing and lets everyone else do the same, it's unclear why we even need the umbrella term "ecology" to refer to all those things.

I'm not the only one who worries about this. Another poll of 247 *Dynamic Ecology* readers asked respondents to rate the field of ecology on a scale of 1 to 5, with 1 meaning that ecology is a coherent, joined-up field, and 5 meaning that ecology is a "disunified mess." The answers skewed toward "disunified mess." The modal response was 4 (chosen by a third of respondents), and only a quarter of respondents chose 1 or 2 (J. Fox 2017).

So the field of ecology definitely has changed in some big ways. But in the eyes of many of us, it remains a mess. Ecologists are diverse in terms of our research goals and approaches, but as a field we're struggling to fully harness that diversity. Struggling to turn diversity of ideas and approaches from a (perceived) problem into a strength. I think we can do better by drawing on lessons from our own study systems. Ecologists have learned a lot about the benefits of diversity in nature. Next I'll review those benefits and suggest that ecologists can learn from them.

2 *The Benefits of Diversity, in Nature and in Ecology*

The first ecological experiment we know of predates Ernst Haeckel's 1866 coining of the term "ecology" by 50 years (Egerton 2013; Hector and Hooper 2002). In 1816, George Sinclair, head gardener of the Duke of Bedford, reported results of an experiment he conducted at Woburn Abbey in England. The experiment compared the yield of hay produced by different mixtures of grasses and herbs growing on different types of soil (fig. 2.1). Some plots were seeded with a single species; others held transplanted turfs containing diverse mixtures of many species. Some plots contained a single type of soil, others contained custom-made mixtures of different types of soil. Sinclair concluded that plots planted with diverse mixtures of species produced more hay, a result that led Charles Darwin (1859, 113) to write in the *Origin of Species*:

> It has been experimentally proved that if a plot of ground be sown with one species of grass, and a similar plot be sown with several distinct genera of grasses, a greater number of plants and a greater weight of dry herbage can thus be raised.

Darwin cited Sinclair's experiment as one line of evidence for his "principle of divergence," which claims that natural selection will favor individuals that produce ecologically divergent offspring. If your offspring can take up different places in the "economy of nature," they will not compete with one another, thereby allowing more of them to survive and reproduce.

Figure 2.1. Plan of the *Hortus Gramineus*, or grass garden, at Woburn Abbey

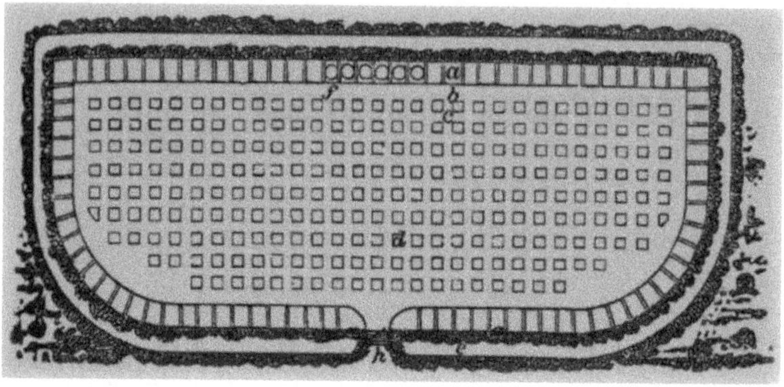

Source: Reproduced from the work of George Sinclair by Hector and Hooper (2002).

Sinclair's results wouldn't convince an ecologist today. A transplanted turf of established grassland will outproduce a freshly seeded plot in the short term, regardless of whether the former contains more species than the latter. But the idea that the collective performance of a group of species often will increase with its diversity has proven durable. There is now a massive biodiversity-ecosystem function (BEF) literature, asking how the level or rate of ecosystem-level properties or "functions" such as total plant biomass, primary productivity, decomposition rate, nutrient retention, and the like, depends on how many species are present and how those species are ecologically differentiated from one another (Tilman, Isbell, and Cowles 2014).

In this chapter, I'll review that literature, focusing on the underlying mechanisms by which more diverse ecosystems can outperform less diverse ones. And I'll suggest that those mechanisms all have scientific analogs: they're all mechanisms by which a scientific field that is diverse in terms of its questions, goals, study systems, and approaches can progress better than a field in which everyone agrees on what question to ask and how to go about answering it.

Selection Effects

To understand selection effects, let's first recall Sinclair's original experiment and how later ecologists improved it.

Starting in the mid-1990s, many ecologists began doing improved versions of Sinclair's experiment. They were interested in how the species richness of a mixture of plants affects its total biomass. To answer that question, you need to experimentally manipulate species richness while holding other variables constant, at least on average. That way, species richness isn't confounded with any other variable, the way it is in Sinclair's experiment. In particular, you don't want to confound species richness (number of species) with species composition (identity of species). You don't want to compare, say, a single-species community of goldenrod with a two-species community of little bluestem and clover. How would you know that any difference in total biomass between the two communities is due to the difference in species richness, as opposed to some difference between, say, goldenrod and little bluestem?

The appropriate experimental design for isolating any effect of species richness is known as a "random draws design." You choose a set of species to use in your experiment; that set is called the "species pool." It has some number of species in it, *S*. You plant replicate plots containing an equal mixture of all *S* species; those are your maximally species-rich plots. You also plant replicate single-species plots, termed "monocultures," of each of the *S* species. And you plant replicates of various mixtures with other species richness levels. Say, some two-species plots, some four-species plots, etc. (The choice of which species richness levels to use is fairly arbitrary and isn't relevant for our purposes.) Each two-species plot comprises an equal mixture of two species chosen at random from the species pool, and similarly for the other species richness levels between 1 and *S*. Hence the name "random draws design." By randomly sampling your two-species mixtures from the population of all possible two-species mixtures, and similarly for the other species richness levels, you ensure (well, make it very likely) that species richness isn't confounded with species identity.

What can happen in a random draws experiment? In particular, why might total plant biomass increase with species richness? One possibility is that each plot will be taken over by the most competitive species planted in it. If more-competitive species produce more biomass, the end result will be that total biomass increases with species richness. This is known as a "selection effect," by analogy with evolution by natural selection (Loreau and Hector 2001; J. Fox 2005). A population evolving via natural selection will come to consist of individuals possessing the phenotypic trait that confers the highest fitness, at the expense of individuals possessing less-fit phenotypes. Analogously, a community exhibiting a selection effect will come to be dominated by the most competitive plant species, at the expense of less competitive species.

For instance, consider an extreme case: an experiment with $S = 10$ species, one of which is a great, highly productive competitor and the other nine of which aren't. Our great competitor takes over any plot it's planted in and produces 900 g of biomass while eliminating all other species. The other nine plant species are all equally poor competitors and equally unproductive, so plots with any number and combination of those nine species produce 100 g of biomass. Total biomass increases with species richness. The average monoculture produces 180 g of biomass ([900 g + 9(100 g)]/10 = 180 g), while a 10-species plot produces 900 g.

"Diversity" here means differential ability: some species are different from others because they're *better* than others at performing the ecosystem function of interest. Diversity in this sense promotes selection effects by providing the "fuel" for selection to "burn." If all plant species were equally productive, there would be no scope for a selection effect to increase primary productivity. And you do need to burn the fuel in order to get the selection effect. If there were no selection—no increase in the frequency of some species at the expense of others—there would be no selection effect, and no increase in total biomass with increasing richness.

Analogously, one benefit of having many different approaches to a scientific task or goal is the ability to select the best approach from among the available options. Chapter 4 provides examples of selection effects in the progress of ecology, including examples in which progress has been hindered by a lack of selection or by selection favoring the wrong approaches.

I am far from the first to suggest that selection effects drive scientific progress. Numerous historians and philosophers of science have argued that science progresses by a process analogous to evolution by natural selection (e.g., Hull 1990). But until fairly recently, they have paid less attention to the other mechanisms by which diversity of scientists' motivations, goals, and approaches can promote scientific progress.

Complementarity Effects

Return again to our imaginary random draws experiment. Selection effects aren't the only reason why species-rich plots might outproduce monocultures on average. If many different species perform the ecosystem function of interest but do so in complementary ways, collectively they will function at a higher level than any one of them could on its own.

"Complementarity effect" is a broad umbrella term. There are various ways species can complement one another so that diverse ecosystems outproduce monocultures. Here are some of them:

- Species are *differentiated* in terms of the resources they are best able to use in order to produce biomass, and/or the times and places at which they are best able to use those resources. So a diverse mixture of multiple species can make fuller use of more resources, at more times and places, than any one species could on its own. For instance, different plant species might differ in rooting depth (Bakker et al. 2021). They might differ in phenology, achieving peak biomass at different times during the growing season (Sánchez-Ochoa et al. 2022). Or they might differ in the specialized pathogens to which they are vulnerable, so that in a diverse mixture of multiple plant species, every species' pathogens remain at low prevalence due to the low density of hosts (Strauss et al. 2024).
- Some species *facilitate* the growth of others (A. Wright et al. 2021). For instance, nitrogen is a limiting resource for plant growth in many terrestrial ecosystems. Nitrogen-fixing legumes can increase nitrogen availability in the soil, making other plant species more productive than if they were growing in a monoculture without legumes.
- Species can *trade resources* with one another. Facilitation of nonlegumes by legumes is a happy accident, from the perspective of the nonlegumes. Legumes fix atmospheric nitrogen in order to aid their own growth. But some of that nitrogen happens to end up in the soil, where it's available to other plant species. In contrast, there are circumstances in which species deliberately

trade resources with one another, increasing the productivity of both species. For instance, nitrogen-fixing legumes don't actually fix atmospheric nitrogen themselves. Rather, specialized endosymbiotic bacteria living in nodules within the plant roots fix the nitrogen. The host plant receives some of this nitrogen, in exchange for sugars and other carbon compounds the plant produces via photosynthesis (e.g., Keller and Lau 2018).

A common thread unifies all these different examples of complementarity effects. Complementarity effects occur when different species, are, well, different. Not different in the sense that some are *better* than others, as with selection effects. Just different. Analogously, ecologists using a complementary mixture of different approaches (e.g., collecting observational field data, conducting experiments, building mathematical models) often can answer scientific questions that would be intractable if fewer approaches were used. Chapter 3 provides examples.

Diverse Tools for Diverse Jobs

So far we've been focusing on how biodiversity affects a single ecosystem function: primary productivity. But of course, different species do all sorts of different things. Many of these things can be thought of as ecosystem functions in their own right, not just means to the end of increasing primary production (the ecosystem function on which we've focused so far). If different species perform different ecosystem functions, collectively they will perform more functions than any one of them could on its own. Analogously, different ecologists pursuing different goals can accomplish more than if we all were to pursue the same goal. Chapter 5 provides examples.

3 *Complementarity*

In BEF experiments, complementarity effects are by far the most important reason why ecosystem functions like primary productivity increase with species richness (Hong et al. 2022). Analogously, complementarity is perhaps the most obvious way a diversity of ideas and approaches can accelerate scientific progress. If I asked you to tell me about the first example that comes to mind of diversity and ideas and approaches accelerating scientific progress, you'd probably think of an example of complementary ideas and approaches.

But although complementarity *in general* is the most obvious way a diversity of ideas and approaches can accelerate scientific progress, *specific examples* of complementarity in science often aren't obvious at all. They often involve a lot of creativity on the part of ecological researchers. Pursuing complementary research approaches often requires creative thinking. For just about any scientific goal you might have, there are probably more complementary resources that would help achieve that goal than you probably realize. You can draw on them, if only you cast a wide net, keep an open mind, and seek out collaborators.

Different Kinds of Complementarity

"Complementarity" is a broad term. Just as there are various mechanisms by which complementarity effects arise in studies of biodiversity and ecosystem function, there are various ways different scientific ideas, goals, and approaches can complement one another:

Complementary strengths and weaknesses. A plant community can be more productive if different plants have, say, different rooting depths or different phenologies. Each plant species grows best at times or places at which other species don't grow well, or at all, thereby making a diverse plant community more productive than any one species could be on its own. Analogously, the weaknesses or limitations of one scientific research approach can be covered up by the strengths of a different approach; the two approaches together outperform either approach on its own.

Facilitation and resource exchange. Some species can increase the productivity of others, either accidentally or as a by-product of deliberate resource exchange. Analogously, one scientific approach can complement another by making it work better than it would on its own—either accidentally or deliberately.

Other forms of complementarity that lack ecological analogs. I'm thinking in particular of complementarity among all the various lines of evidence and argument required to establish many scientific claims. A scientific claim like "These competing marine algae coexist by partitioning the light spectrum (Stomp et al. 2007)" or "Genetic rescue will restore the tiny Florida panther population to viability by reducing inbreeding depression (Onorato et al. 2024)" often is best thought of as a whole bunch of subclaims that *all* need to be true in order for us to reliably infer that the claim as a whole is true. Further, different subclaims often have to be established using different scientific approaches. The best analogs here aren't from ecology. This kind of complementarity is more analogous to, say, solving a crossword puzzle: you haven't solved it until you've filled in every word (Haack 2000). Another analogy is to the organs in your body: they all have their own essential functions, which they all need to perform for you to survive and thrive.

Complementary Strengths and Weaknesses: Different Sampling Methods

First, let's talk about scientific complementarity in the sense of different methods with different, complementary strengths and weaknesses. Analogous to complementarity among plant species with different rooting depths or phenologies.

One of the most important descriptive tasks in ecology is figuring out when and where different species are found, at what abundance. Trouble is, intra- and interspecific variation in habitat, body size, coloration, behavior, and other factors means that different sampling methods work best for different species. Fortunately, that problem contains the seed of its own solution. Different sampling methods complement one another: each can detect some species, at some places or times, that other methods would not. Using multiple sampling methods therefore allows more complete and accurate sampling of species richness, composition, and abundance than would be possible with any one sampling method on its own ("triangulation"; Campbell et al. 2018). Examples include complementarity of traditional sampling methods and DNA metabarcoding for sampling aquatic microalgae and macroinvertebrates (Keck et al. 2022; fig. 3.1), and the complementarity of professional sampling and citizen science observations (Ward-Fear et al. 2019; Krabbenhoft and Kashian 2020; Dimson et al. 2023).

Figure 3.1. DNA metabarcoding complements traditional methods for sampling microalgae (a) and aquatic macroinvertebrates (b)

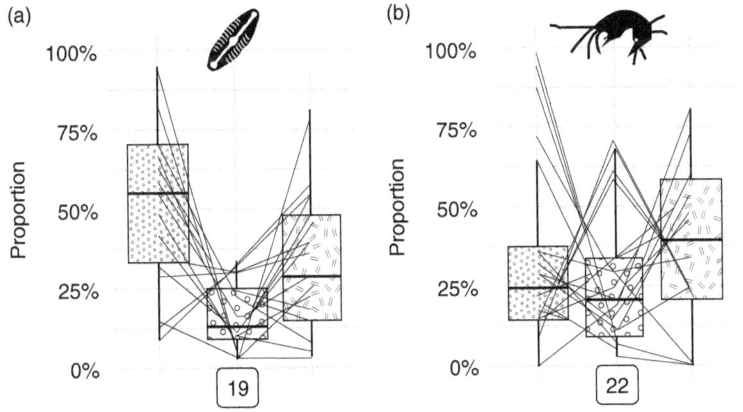

Box plots of the percentage of species detected by traditional methods (*left box plot in each panel*), DNA metabarcoding (*right box plot in each panel*), or both (*center box plot in each panel*). Each box plot gives the distribution of data from many different studies. Lines connect values from individual studies. Relatively few species are detected by both DNA metabarcoding and traditional methods, indicating that each method detects some species that the other does not.

Source: Modified from Keck et al. (2022).

Some of my favorite examples of complementarity among different sampling methods involve not new technology, but old technology—such as the "technology" of people remembering what they and others have seen and done. Here's an outstanding example, pointed out to me by Jarrett Byrnes.

Anne Salomon, Nick Tanape Sr., and Henry Huntington (2007) wanted to explain recent declines in the harvests of black leather chitons (*Katharina tunicata*) from sites on the outer Kenai Peninsula in Alaska. Black leather chitons live in rocky intertidal sites in this area. The chitons are an important subsistence fishery for Sugpiaq (Chugach Alutiiq) Alaska Native people who live in two villages in the area. Black chitons are the most common invertebrate in landings records from the village subsistence fisheries spanning 1987 to 2003. But black chiton landings declined over the time period covered by landings records.

In order to explain this decline, Salomon, Tanape, and Huntington used standard ecological methods to survey 11 field sites. They collected data on chiton abundance and size structure, and on variables that might affect abundance (wave action, presence/absence of birds and sea otters that prey on chitons, etc.). They also gave Native villagers who harvest chitons a questionnaire to quantify harvesting effort at each site. But information on current chiton abundance, current harvesting effort, current otter presence/absence, and so on doesn't provide the full picture. It doesn't reveal the past events that ultimately set the stage for recent chiton declines.

Which is where complementary sampling methods come in. Salomon, Tanape, and Huntington drew on three other complementary sources of information: archaeological data, historical records, and interviews with tribal elders. Archaeological excavation of a local midden site at Port Graham revealed that in prehistoric times (~1300–1500 CE), the Native invertebrate harvest was dominated by large predatory whelks (*Neptunea* spp.). In early historic times (20th century), *Neptunea* spp. made up a modestly smaller proportion of the invertebrate harvest. Smaller predatory whelks, butter clams, and other smaller-bodied species made up a modestly increased proportion of the invertebrate harvest. Black chitons contributed only a very small fraction of the invertebrate harvest in prehistoric or early historic times. Historical records show that Port Graham was established as a permanent settlement between 1909 and 1912 to support the growing shipping and fishing industries. The Sugpiaq people became more economically intertwined with the fishing industry. They gave up their previous seminomadic lifestyle and settled in permanent villages. Over the years, the villagers acquired new technologies, such as outboard motors and freezers. The recollections of Sugpiaq elders document those changes in lifestyle and technology and their consequences for the subsistence fishery. The primary consequence was a series of declines in marine invertebrates, beginning in the early 1960s, as the local sea otter population recovered, and new technology improved the efficiency of subsistence harvesting. Larger and more-preferred invertebrate species declined first, prompting subsistence harvesters to switch to smaller and less-preferred species that declined in turn. Localized declines in black chitons in the late 20th century were merely the most recent in a sequence of declines.

Facilitation and Resource Exchange: Indirect Evidence

Next, let's talk about complementarity in science that's analogous to facilitation and resource exchange in BEF experiments. There are some species that improve the survival, growth, and functioning of other species. Analogously, there are scientific research approaches and lines of evidence that don't merely compensate for one another's weaknesses, but actually reduce or eliminate one another's weaknesses, so that each approach or line of evidence is actually improved compared to how it would perform on its own.

There are plenty of examples of intentional facilitation in ecological research—researchers deliberately using one research approach or line of evidence to improve some other approach or line of evidence. But I want you to appreciate just how common complementarity is in scientific research, so I'm going to discuss examples of accidental facilitation. After all, most facilitation in BEF experiments is accidental. In a BEF experiment, nitrogen-fixing plants can facilitate the growth of nonlegumes, even though that's not something the nitrogen fixers are "trying" to do, or have been selected by evolution to do. Facilitation of nonlegumes by legumes is just a happy accident, from the perspective of the nonlegumes. Analogously, different ecological studies can complement and improve one another, even if they weren't intended to do so.

They can complement and improve one another even if they were originally thought to be antagonistic rather than complementary. Indeed, different studies can complement and improve one another even if they have no obvious scientific relationship to one another at all!

In all likelihood, you have used complementary lines of evidence that improve one another, quite possibly without even realizing you were doing so. You've done it if you've ever run a linear regression. Efron (2010) provides an excellent primer on linear regression as a form of what he calls "indirect evidence." "Indirect evidence" is an example of accidental facilitation in my parlance, but I'll stick with Efron's term in order to summarize and extend his argument.

Any linear regression would work as an example. I'll use Efron's (2010) own example (fig. 3.2). The kidney function of healthy volunteer donors declines on average with donor age. The plotted measure of kidney function is derived from a series of medical tests. A new donor, age 55, volunteers to donate a kidney. What's our best estimate of the new donor's kidney function?

Figure 3.2. A measure of kidney function versus age in years for 157 healthy volunteer kidney donors

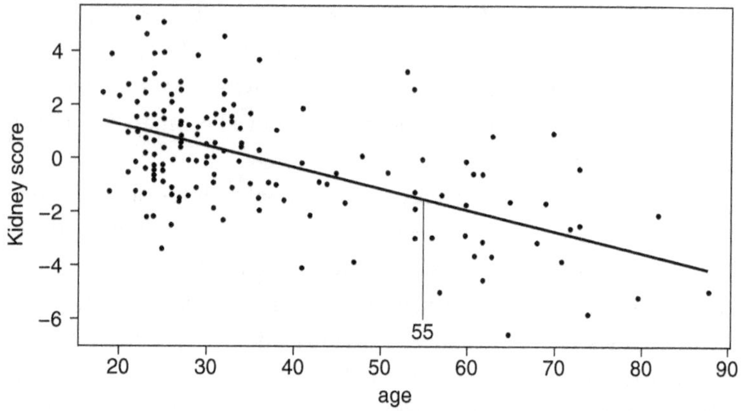

The negatively sloped least squares regression line indicates that estimated mean kidney function declines with age. A new 55-year-old donor volunteers. Our best estimate of the new donor's kidney function is not the observed kidney function of the only 55-year-old patient in the dataset but rather the regression-based estimate of mean kidney function at age 55.

Source: Modified from Efron (2010).

One way to answer that question is to rely solely on "direct evidence"—that is, evidence from other donors who are exactly like our new donor in relevant respects. Here, the "relevant respect" is age. Our dataset contains one observation, of a 55-year-old donor who had a kidney function score of -0.01. So we could estimate that our new donor is most likely to have a kidney function score of -0.01.

But that estimate would be bad, because it ignores all the indirect evidence at our disposal. We have data from 156 donors of other ages. That evidence remains relevant despite the fact that none of those other patients is 55 years old, because we can use linear regression to estimate how mean kidney function depends on age. Our linear regression estimates that mean kidney function of 55-year-olds is -1.46. Importantly, that estimated value depends on every observation in the dataset. As statistician John Tukey put it, we're "borrowing strength" from observations of patients who *aren't* 55 years old, to improve our estimate of the mean kidney function of 55-year-olds (Brillinger 2002, 1557). Observations of kidney function in patients of different ages thus facilitate one another. They all can be used together to improve the accuracy and precision of estimated mean kidney function at any age, beyond what would be possible if we relied solely on direct evidence.

OK, so you can think of linear regression as a form of indirect evidence, a kind of facilitation. The indirect evidence improves the direct evidence, like a nitrogen-fixing legume improving the growth of a non-legume. So what? Why is it helpful to think of linear regression that way? For two reasons, I think. First, it helps you see indirect evidence as something normal, even obvious, rather than something fishy or overly clever, that you ought to be reflexively suspicious of. Which leads to the second reason: seeing indirect evidence as just normal evidence helps you recognize opportunities to cast a wide net—to draw on indirect evidence that you might have overlooked otherwise.

Here's one of my favorite examples: V. Smith et al. (2005). The authors wanted to describe the species-area curve for phytoplankton. At the time, species-area curves had been well described for many terrestrial taxa in many places, but not for aquatic species as small as phytoplankton (Rosenzweig 1995). And many hypotheses had been suggested to explain why species-area curves look as they do. But without knowing what the species-area curve looks like for phytoplankton, we couldn't do more than speculate which of those hypotheses might or might not apply to phytoplankton.

Species-area curves for terrestrial species often are well described by a power law: $S = bA^z$, where S is the number of species in a habitat of area A, and b and z are constants. Log-transforming both sides produces the linear relationship $\log(S) = \log(b) + z\log(A)$, where z is now a slope that can be estimated via linear regression of $\log(S)$ on $\log(A)$. To get a precise, accurate estimate of z, and to be able to detect any nonlinearity in the relationship between $\log(S)$ and $\log(A)$, you want to sample the largest range of areas possible. Which is why, when the authors of V. Smith et al. (2005) compiled data on phytoplankton species richness and area, they didn't just include data from natural water bodies, from small ponds up through lakes to ocean basins. They also included data from artificial water bodies: microcosms and mesocosms. Microcosms and mesocosms range from small artificial ponds made from cattle-watering tanks, with surface areas of a few square meters, down to tiny flow-through cultures known as chemostats, containing less than a liter of water, with surface areas measured in square centimeters.

I think many ecologists would not have included microcosms and mesocosms in the same dataset as natural water bodies. After all, they're different in so many ways! Many aspects of the microcosm or mesocosm environment are controlled, and sometimes manipulated, by the investigator—everything from water temperature to concentrations of key nutrients such as nitrogen and phosphorus. Many microcosms and mesocosms lack zooplankton, fish, and other species that ordinarily interact with phytoplankton in nature in various ways (e.g., by consuming phytoplankton, or consuming the consumers of phytoplankton). The investigator running a microcosm or mesocosm experiment often also controls the phytoplankton themselves. The maximum possible species richness for a chemostat is set by the number of species the investigator adds (not all of which will necessarily persist). Prominent aquatic ecosystem ecologist Steve Carpenter (1996) argued that the artificiality of microcosms and mesocosms makes them fundamentally incomparable to natural water bodies. Prominent terrestrial ecologist Charles Krebs argued the same; Krebs (2015b) compares the artificiality of microcosms to Volkswagen cars that were illegally rigged to perform differently under emissions testing than under natural driving conditions. According to these ecologists and others, microcosms and mesocosms are just different from nature.

To which: yes, microcosms and nature are different—just as, say, 37- and 55-year-old kidney donors are different. Have a look at figure 3.3, modified from V. Smith et al. (2005).

The phytoplankton species-area curve for microcosms and mesocosms turns out to be statistically indistinguishable from that of natural water bodies, in terms of both the slope and the intercept. And not because of lack of statistical power to detect any differences either—the regressions for each type of water body separately (fig. 3.3, solid lines) are both fairly close to the regression for the pooled data (fig. 3.3, dashed line).

The results of V. Smith et al. (2005) actually illustrate complementarity in both the sense of complementary strengths and weaknesses, and in the sense of facilitation. Complementary strengths and weaknesses: Natural water bodies and artificial water bodies differ in size. Each covers a range of areas that the other mostly doesn't. Together,

Figure 3.3. Phytoplankton exhibit similar species-area curves in microcosms and mesocosms and in natural water bodies

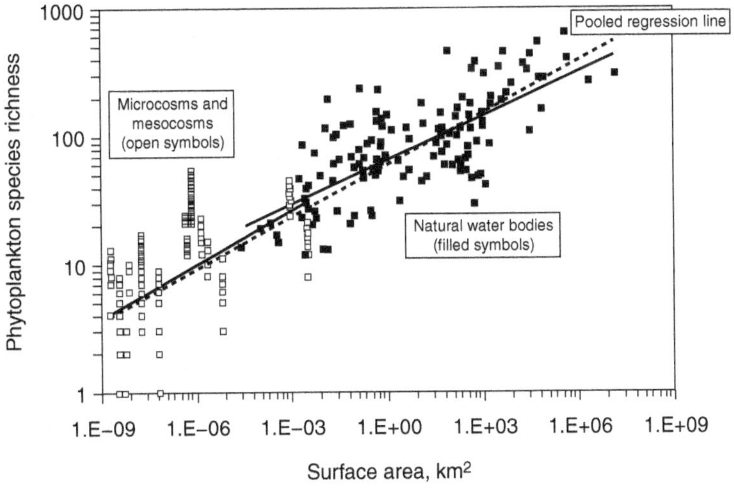

Open and filled symbols give data for microcosms and mesocosms, and natural water bodies, respectively. Solid lines are regression lines for each type of water body. The dashed line gives a linear regression on the pooled data.

Source: Modified from V. Smith et al. (2005).

they cover a larger size range than either could on its own, confirming that the phytoplankton species-area curve is linear (on a log-log plot) over the entire size range of water bodies. This is interesting because some hypotheses, and data from some terrestrial taxa, suggest that species-area curves are, or should be, triphasic, meaning that the slope z should differ for small, intermediate, and large areas (Palmer 2007; Storch 2016). That's complementarity in the sense of complementary strengths and weaknesses, discussed earlier in this chapter. Facilitation: data from natural and artificial water bodies are complementary in Efron's (2010) sense of indirect evidence. Microcosms and mesocosms provide indirect evidence about the species-area curve for natural water bodies, and vice versa. Our best estimate of mean phytoplankton species richness in a water body of any given size is slightly different—and slightly better—than it would have been had the authors of V. Smith et al. (2005) considered only natural water bodies. That's complementarity in the sense of one line of evidence improving another, analogous to a legume facilitating the growth of a nonlegume.

The idea of indirect evidence can be stretched even further. In V. Smith et al. (2005), measurements of the same variables (phytoplankton species richness and water-body surface area) are mutually informative, whether they're taken from artificial or natural water bodies. It turns out that measurements, or estimates, of *different* variables also can be mutually informative. Estimates of, say, the effects of wild bees on crop pollination can be used to improve estimates of, say, the effect of predators on larval fish survival, and vice versa. Yes, really! Explaining why would take long enough to test your patience, so I won't try. Check out J. Fox (2022) if you're curious.

Complementary Lines of Evidence Addressing Different Subclaims: Disease Dynamics in African Buffalo

Finally, let's turn our attention to a third form of scientific complementarity, one that doesn't really have an ecological analog—at least, not one I can think of! Many scientific claims about the world actually comprise a bunch of subclaims. The subclaims all have to be true in order for us to reliably infer that the claim as a whole is completely true. Often, checking different subclaims demands different approaches and lines of evidence. These different approaches and lines of evidence are complementary to one another because they speak to different subclaims.

Here's an outstanding example on disease dynamics in African buffalo, thanks to Vanessa Ezenwa and her colleagues. Multiple parasites and/or pathogens often infect the same individual host. Parasites and pathogens infecting the same host can affect one another's survival and reproduction, via the host's immune system. That's because the host's immune response to one parasite or pathogen can alter its immune response to a co-infecting parasite or pathogen. Helminths are parasitic worms. They are common extracellular macroparasites of vertebrates. Laboratory studies of mice show that infection with helminths induces upregulation of the vertebrate T-helper type 2 (Th2) immune response. The Th2 immune response enhances production of molecules such as interleukin-4 (IL-4) and eosinophils that attack helminths. But a side effect of helminth infection is downregulation of the T-helper type 1 (Th1) immune response. The Th1 immune response produces molecules such as interferon gamma (IFNG) that attack intracellular microparasites such as viruses and bacteria. Further, just as the Th2 response interferes with the Th1 response, the Th1 response interferes with the Th2 response. Laboratory studies of mice indicate that upregulation of the Th1 response causes downregulation of the Th2 response. Laboratory studies of mice also show that, as a consequence of these immune responses, infection with helminths makes mice more susceptible to microparasite infection, and can increase the duration and severity of microparasite infection.

Based on that background information, you'd expect that individual vertebrate hosts in nature that can mount a stronger immune response to helminths would be more susceptible to microparasite infection, and would experience more-intense and longer-lasting microparasite infections if infected. The consequences of this for population-level microparasite prevalence are less clear, however. On the one hand, previous or ongoing exposure to helminths might enhance the duration and intensity of intracellular microparasite infections. This would enhance microparasite transmission, making microparasites better able to invade and persist in host populations previously unexposed to microparasites. On the other hand, hosts unable to mount a Th1 response to microparasites—for instance, due to co-infection with helminths—might die quickly. Quick death of co-infected hosts would remove them from the population, possibly preventing both helminths and microparasites from spreading to other host individuals.

Ezenwa and colleagues set out to test this hypothesis in African buffalo (*Syncerus caffer*). They studied buffalo in a national park in South Africa. These buffalo are host to both gastrointestinal helminths and (since 1986) *Mycobacterium bovis*, the bacterium that causes bovine tuberculosis (TB). Perhaps bovine TB was able to establish only because these buffalo had previous (and ongoing) exposure to helminths?

Now, stop and think for a moment about *everything* that has to be true if this hypothesis is correct. Write down a list! Once you're done, compare your list to that of Ezenwa and colleagues (Jolles et al. 2008; Ezenwa, Etienne, et al. 2010):

- Helminth infection should produce a Th2 response.
- TB infection should produce a Th1 response.
- All else being equal (e.g., host age, sex), individual hosts that mount stronger Th2 responses should mount weaker Th1 responses, and vice versa.
- All else being equal, co-infected hosts should be in poorer health than singly infected hosts, which should be in poorer health than uninfected hosts.
- Co-infected hosts should experience high mortality.
- High mortality of co-infected hosts should have multiple knock-on effects. At the individual level, it should remove individuals with both intense helminth infections and intense TB infections. That is, any individuals with extremely high helminth burdens should be TB-free, because otherwise they'd be dead. At the herd level, high mortality of co-infected hosts should generate a negative correlation across herds between helminth prevalence and TB prevalence.
- Treating helminth infections with anti-helminth drugs should increase the Th1 response, thereby reducing susceptibility to and duration of TB infection.
- The net result of all these effects could be that helminths promote TB invasion into previously TB-free buffalo populations. But whether or not helminths promote TB invasion will depend on the relative magnitudes of different effects (e.g., effect on transmission rate of TB vs. effect on mortality rate of co-infected individuals).

Ezenwa and colleagues used a diverse mixture of research approaches to check every item on that bulleted list. (Here I'll just summarize some of the results; see Jolles et al. 2008 and Ezenwa, Etienne, et al. 2010 for the full results.) Buffalo blood samples indicate that individuals that mount a strong Th2 response mount a weak Th1 response, and vice versa. Co-infected individuals exhibit the worst body condition on average, while uninfected individuals exhibit the best condition on average, with singly infected individuals being intermediate. Co-infected individuals with an extremely high worm burden aren't observed, presumably because they're dead. Herds with higher observed helminth prevalence exhibit lower TB prevalence and vice versa, presumably because co-infected individuals die quickly. Experimental helminth removal (deworming) improves host body condition and the Th1 immune response against microparasites. There is genetic variation for helminth resistance at the host BL4 locus. Host individuals carrying the low-resistance allele BL4-169 carried higher-average helminth burdens than other individuals, and also mounted a stronger Th1 response against microparasites than other individuals. Finally, in a mathematical model of disease dynamics parameterized from the observational and experimental data, TB can invade a TB-free buffalo population only in the presence of helminths.

There's a twist ending to this story, though. Extensive and impressive as that bulleted list is, there's one item missing from it. If you want to confirm beyond any reasonable doubt that helminths promote invasion and persistence of TB, you can do one more check. You can experimentally deworm buffalo and then see whether TB goes extinct. To their great credit, Vanessa Ezenwa and Anna Jolles (2015) did that experiment, on a different buffalo population in a different South African national park (Kruger National Park). They got a shock: helminth removal actually *massively increased* TB's estimated ability to invade a previously unexposed buffalo population (fig. 3.4). That's the precise opposite of what one would expect, based on all their previous work on buffalo, and based on previous laboratory work in mice! The explanation seems to be differences in transmission and mortality rates between buffalo populations. In the Kruger buffalo population, the relative magnitudes of the effects of helminth infection on TB susceptibility transmission versus effects on TB mortality apparently are very different than they are elsewhere in South Africa.

Figure 3.4. Experimental helminth removal increased the estimated ability of TB to invade a previously TB-free buffalo population

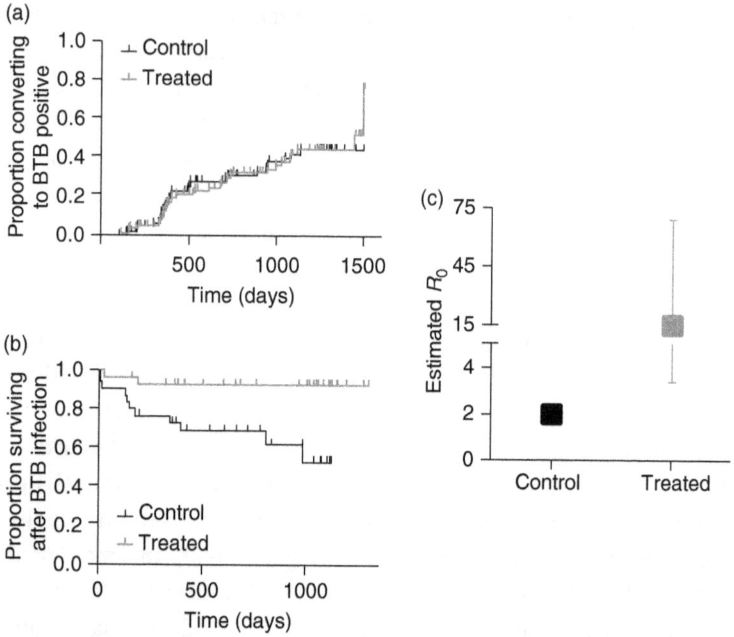

Experimental removal of helminths via deworming had no effect on TB prevalence (a), but increased buffalo survival after TB infection (b). This increased survival following TB infection improved the estimated ability of TB to spread in a previously TB-free buffalo population (c; higher R_0 indicates increased ability of a pathogen to spread in a previously pathogen-free host population).

Source: Modified from Ezenwa and Jolles (2015).

Conclusions

The surprising results of Ezenwa and Jolles (2015) illustrate the power and value of complementary lines of evidence in ecology, not the weaknesses or limitations of complementary lines of evidence. After all, in the absence of all the complementary lines of evidence from Jolles et al. (2008) and Ezenwa, Etienne, et al. (2010), we wouldn't even realize just how surprising the results of Ezenwa and Jolles (2015) are, never mind being able to (tentatively) explain why those surprising results occurred. Conversely, if Ezenwa and Jolles hadn't done their experiment, we might well have been overconfident in our other lines of evidence regarding how co-infections affect TB dynamics in buffalo.

More broadly, the results of Ezenwa and Jolles (2015) illustrate just how variable and heterogeneous ecological studies tend to be (Senior et al. 2016; Bebout and Fox 2024). In order to have any hope of cutting that variability down to size and explaining where it comes from, we're going to need to draw on all the information we can get. Which is what complementarity in ecological research is all about, I think. It's just *so* difficult to learn *anything* with much confidence in ecology. Especially anything that generalizes to other times/places/species/whatever. Knowing that it's so difficult should motivate us to rise to the challenge—to find creative ways attack our scientific goals from all angles, drumming up every piece of complementary evidence we can.

4 *Selection*

Taking full advantage of diversity sometimes means getting rid of it.

That's how selection effects work in BEF experiments. In an initially diverse plot, the more-productive plant species increase in relative abundance at the expense of the less-productive plant species, thereby raising the productivity of the plot as a whole. Diversity—here, variation in plant productivity—is merely "fuel" for selection to "burn." And it does need to burn, if it's to be of any functional benefit. Insofar as the initial diversity is maintained over the long term, that's a failure of selection. Less-productive species create a drag on total productivity. Productivity would be higher if those species were outcompeted by more-productive species.

Analogously, sometimes in science we want to have a diversity of methods or approaches, purely so that we can select the best one. The others get discarded. We don't want to stick with a suboptimal method or approach if better options are available.

Ecologists select among diverse methods all the time. As one small illustration, we have an entire journal, *Methods in Ecology and Evolution*, filled with papers that propose new methods for performing all sorts of technical tasks. Many of those papers argue that those new methods improve on previous methods and so should replace previous methods.

Model selection provides another familiar example of selection in ecology—it's right there in the name (Burnham and Anderson 2003; Tredennick et al. 2021)! Model selection involves choosing the best statistical model for the data from among a set of candidate models.

The "best" model is the one that best meets some criterion, such as lowest AIC score (an estimate of the model's ability to predict new observations sampled from the same statistical population from which previous observations were sampled; Burnham and Anderson 2003; Tredennick et al. 2021).

Selecting among alternative scientific hypotheses is a third familiar example of selection in ecology and other scientific fields—albeit not one that ecologists often attempt (Betini, Avgar, and Fryxell 2017; Betts et al. 2021).

But selection in ecology isn't limited to selecting the best statistical model, the best method to conduct some technical task, or the best alternative hypothesis. You can also think about "selection" in scientific research in a broader and looser sense. For instance, think of the "file-drawer problem"—the fact that many scientific studies go unpublished, and so get left in a metaphorical file drawer (or these days, on a metaphorical server). As we'll see later in this chapter, unpublished studies are a nonrandom sample of all studies (Yang, Hillebrand, et al. 2022; Yang, Sánchez-Tójar, et al. 2023). That is, they're selected: study authors select which studies to submit for publication, and reviewers and editors select which submissions get published.

Selection, in the broad sense of "picking and choosing among options, based on some criteria," also happens within published studies. Empirical ecology papers ordinarily report numerous results—statistical analyses of each of several dependent variables, different analyses of the same variable, and so on. As an author, which results do you select to highlight as the main results in your title and abstract? Conversely, which results do you bury in online appendixes? Scientific research isn't finished until it's published. And so it's important for ecologists to think about how well we select among diverse options, both when we're doing the research, and when we're writing it up.

Selection, in the broad sense of "picking and choosing among options, based on some criteria," happens even when only a single option receives conscious consideration. If I need to drive a screw into a piece of wood, obviously I'll select a screwdriver for the job. That's the right choice for the job, both in the comparative sense of "it'll do the job better than other tools," and in the noncomparative sense of "it'll do the job, period." A screwdriver is the right choice even if I never consciously consider any other option for driving the screw. Conversely, if for some strange reason I try to drive a screw with a hammer, that would be a bad choice. A hammer is the wrong tool for the job, both compared to a screwdriver and in absolute terms. A hammer won't get the job done. It's a bad choice even if I never consider using a screwdriver. Heck, it's a bad choice even if I don't even have a screwdriver. Rather than wasting my time trying to drive a screw with a hammer, I should go do something else instead.

Selection isn't just something that individual researchers do. You can even think about selection at the level of what sort of research the entire field of ecology chooses to pursue, or not pursue. For instance, recall from chapter 2 that ecology as a whole has selected for research on applied topics, starting in the late 1980s (Anderson et al. 2021). Selection for research on applied topics has come at the expense of research on older fundamental topics like competition, predation, and life-history theory. The field as a whole has "voted with its feet" for research on global change, conservation, invasive species, and restoration rather than research on older fundamental topics. Or think of criticisms of ecology journals and funding agencies for publishing and funding novel studies, rather than replications of previous studies (e.g., H. Fraser, Barnett, et al. 2020). The choice to publish and fund novel studies rather than replications of previous studies is a kind of selection.

Every aspect of doing science thus involves constant selection from among diverse options, by individual scientists and by scientists collectively. Some of that selection will be successful, but some of it won't be. Selection in science is often challenging; it involves balancing multiple desiderata, without a crystal ball. It's not always obvious which method or research approach will do the job best, or do it at all.

In this chapter, I'll discuss a range of examples of selection in ecology:

- selecting from among alternative scientific hypotheses
- selecting which results to publish
- selecting which results to highlight in a published paper
- selecting from among approaches to infer process from pattern

I chose these examples to illustrate three different ways selection can go wrong in ecological research:

Failure to select. Imagine a diverse mixture of plant species that produce different amounts of biomass in monoculture. But in mixture, none of those species increases or decreases from its initial relative abundance, so there's no selection effect. There's fuel for selection—the plant species vary in their productivity—but no selection. The fuel doesn't get burned. Or perhaps some species do increase in relative abundance while others decrease, but the increases and decreases are driven by chance events, and so are independent of species productivity. There's no selection effect in that case either. Analogously, in scientific research it can be hard to tell which option is best, in which case, one option might be preferred for arbitrary reasons. Or researchers who find it difficult to choose the best option, or who worry about the costs of making a poor choice, might avoid making any choice at all.

Selecting for the wrong things. This is analogous to what are known as "negative selection effects" in BEF experiments (Jiang 2007). A negative selection effect arises when a diverse mixture of plant species comes to be dominated by the *least*-productive species, at the expense of more-productive species. Negative selection effects cause diverse mixtures to be *less* productive than monocultures are, on average. The ecological processes that determine species' relative abundances in mixture don't select for productivity, but instead for some other attribute that's negatively associated with productivity. Analogously, scientific researchers choosing from among different options often have to weigh many desiderata that can't all be satisfied at once. Which creates the risk that they'll overweight some desiderata and underweight others, and so end up selecting an option that seems reasonable at the time but turns out to be regrettable.

Lack of variation to select from. In a BEF experiment, a monoculture of a single species doesn't allow any scope for a selection effect. Analogously, in scientific research you can't select from among different options if you have only one option (e.g., no alternative hypotheses).

In this chapter, I'll discuss successful and unsuccessful examples of selection in ecological research. We'll start with a case of failure to select: not selecting from among alternative scientific hypotheses. Ecologists rarely even try to select from among alternative scientific hypotheses, and sometimes they fail when they do try. Why is that? Then we'll contrast failure to select from among alternative scientific hypotheses with a comparatively successful example of selection: the peer-review system, by which ecologists select which papers to publish in which journals. The comparative success of selective peer review is informative, because the same factors that (purportedly) make selecting from among alternative scientific hypotheses challenging in ecology also make selective peer review challenging. Next we'll turn our attention from failure to select, to negative selection effects: three examples of ecologists repeatedly selecting for (what turned out to be) the wrong things. We'll discuss authors selecting which results to publish, in biased ways. We'll discuss authors selecting which results to emphasize in their published papers, in biased ways. And we'll discuss ecologists repeatedly selecting ineffective research approaches for inferring "process from pattern." There's a common thread tying together all these examples of negative selection effects: trade-offs. Scientists sometimes select for (what turn out to be) the wrong things, because there are many desiderata in scientific research. Selecting for one desideratum often means sacrificing another. It's hard to choose which sacrifices to make, and sometimes we get it wrong. Finally, lack of variation to select from is a bit of a special case, so we'll defer discussion of that until chapter 6.

Before we dive in, I want to emphasize that I'm not trying to show that ecologists, in general, are bad at selection in general. Rather, I'm merely trying to give you a search image for some ways in which selection in our field can go wrong, without drawing any conclusions as to how often it goes wrong. I have no idea how good ecologists, in general, are at selection in general. My hope is merely that studying these examples can help us improve.

Failure to Select: Not Selecting from among Alternative Scientific Hypotheses

Let's first consider failure to select the best option from among those available, analogous to an initially diverse mixture of plants in which the more-productive species don't outcompete the less-productive ones.

Selecting from among alternative scientific hypotheses is perhaps the most familiar example of selection in science. The classic argument for how to test alternative scientific hypotheses is due to Platt (1964). John Platt argued that scientists ought to line up all the alternative hypotheses that might be true about whatever it is they're studying, and then design a series of "decisive" experiments that would distinguish among those hypotheses. For instance, if hypothesis 1 predicts that *x* will occur in experiment *y*, while hypothesis 2 predicts not-*x*, then experiment *y* is a decisive experiment for purposes of distinguishing between hypotheses 1 and 2. Platt (1964) called this approach "strong inference." It's a methodical approach to selection—you line up all the possibilities, then eliminate them until you're left with the one that's true (or at least, consistent with all your experiments).

"Strong inference" sounds great, but ecologists rarely attempt it (Betini, Avgar, and Fryxell 2017; Betts et al. 2021). Arguably, ecologists rarely attempt strong inference, because it's rarely feasible in ecology (Quinn and Dunham 1983). Platt was a molecular biologist running highly controlled, highly repeatable lab experiments—indeed, *so* controlled and *so* repeatable that if you needed statistics to infer whether the treatment and control groups truly differed, that meant you'd done the experiment sloppily and should do it again more carefully! Further, the alternative hypotheses of interest in molecular biology at the time were mutually exclusive, and either right or wrong. In contrast to 1960s molecular biologists, ecologists have to deal with many sources of variation, including but not limited to sampling error. Further, alternative scientific hypotheses in ecology often aren't mutually exclusive, often interact in complex ways, and are often a matter of degree rather than being totally true or totally false. Basically (it's claimed), the bits of nature that ecologists study are just too messy and complicated for us to select from among alternative hypotheses.

No doubt there's something to that argument. Certainly it resonates with the BEF analogy: selection effects in BEF experiments tend to be rather weak, basically because community ecology is messy and complicated. Plants need various resources obtainable from various sources; they're attacked by various natural enemies; their growth and survival depend on various abiotic variables that fluctuate on various timescales; and so on. The messiness and complexity of community ecology creates plenty of opportunities for complementarity effects in BEF experiments, but it prevents strong selection effects. However, I worry there's not quite as much to that argument as many ecologists seem to think. To be clear, I don't think ecologists should, or could, test alternative scientific hypotheses in most or all of our papers. I can't even put a number on exactly how much hypothesis testing we could or should do. I just worry that we're missing some unknown number of opportunities—that we could, and should, be testing alternative scientific hypotheses more often than we currently do.

One reason to worry that we're missing some opportunities to test alternative scientific hypotheses is just the fact that we sometimes *do* test alternative scientific hypotheses. Arguing that ecologists can't test alternative hypotheses because ecology is messy and complicated, proves too much. *All* ecology is messy and complicated—at least, if you insist on looking at it that way (see chapter 7). But yet, some ecology papers do manage to test alternative scientific hypotheses, and not because they concern the rare ecological topics that are inherently "neat and simple" rather than "messy and complicated." So "Ecology is messy and complicated" doesn't work as a blanket excuse for not testing alternative scientific hypotheses in ecology.

For what it's worth, I'm not the only one who thinks ecologists are missing some opportunities to test alternative scientific hypotheses. Here are a bunch of examples of ecologists testing alternative scientific hypotheses, and/or calling on other ecologists to do so more often and explaining how they could do so.

- The authors of Kendall, Briggs, et al. (1999) performed strong inference about the causes of population cycles in various species, although they used other means besides decisive experiments to select among their alternative hypotheses.
- McCauley et al. (1999) is the culmination of series of observations and experiments by Ed McCauley and coauthors to explain why the freshwater zooplankton *Daphnia* spp. and their algal prey don't exhibit the theoretically predicted "paradox of enrichment" (increasingly high-amplitude predator-prey cycles in more resource-rich habitats). The authors tested five alternative

hypotheses for the lack of a paradox of enrichment in *Daphnia*. They managed to do this even though *Daphnia* population dynamics in nature are noisy, and even though the alternative hypotheses aren't mutually exclusive.

- Schluter and McPhail (1992) develops a now-standard checklist of assumptions and predictions that need to be tested in order to demonstrate character displacement (competing species evolving to use different resources). The checklist includes alternative hypotheses that must be ruled out in order to warrant an inference of character displacement. For instance, here's one of the items on the checklist: you have to rule out the possibility that species evolved divergent resource use while living in different locations from one another ("divergence in allopatry") and only later began living in the same locations ("secondary contact"). A few natural and lab-based study systems have now been shown to tick every box on the list. A recent review argues that character displacement researchers should put more effort into ticking every box on the list, rather than settling for ticking only those boxes that are easier to tick (Stuart and Losos 2013). The Schluter and McPhail (1992) checklist is a particularly telling counterexample to the claim that you can't do strong inference in ecology because ecology is "random" or "noisy." Ruling out "random chance" is the first item on the Schluter and McPhail (1992) checklist.
- Sticking with examples from evolutionary ecology: Simons (2011) develops a ranking of different lines of evidence for "bet hedging." "Bet hedging" refers to life-history traits that are favored by natural selection in unpredictable variable environments. Bet-hedging traits reduce the expected number of descendants an individual will leave in the next generation in order to reduce the risk of leaving few or no descendants (say, because you die or fail to reproduce due to a severe drought). Stronger lines of evidence for bet hedging are those that rule out alternative possibilities. Weaker lines of evidence are those that are merely consistent with bet hedging while also being consistent with other possibilities. Stronger lines of evidence are more difficult to produce and so have been produced less often. But they *have* been produced (Simons 2011).

- Ford and Goheen (2015) calls for studies of trophic cascades involving large carnivores to be based on strong inference, and explains in detail how this could be done. Even though large carnivores are infamously difficult to study in controlled replicated experiments at the relevant spatial and temporal scales, and even though the alternative hypotheses aren't mutually exclusive.
- Downes (2010) calls for strong inference in studies of effects of multiple stressors in streams. Notably, Downes (2010) uses the noisiness and complexity of stream ecosystems as an argument *for* strong inference, not an argument against it.
- O'Connor et al. (2015) reviews studies of climate change impacts. The review finds that only a minority stated prior expectations and tested alternative hypotheses about drivers of change. But the fact that some studies did test alternative hypotheses shows that strong inference of climate change impacts is possible. Failure of many climate-impact studies to use strong inference has nothing to do with the messiness or complexity of the problem.
- Dochtermann and Jenkins (2011) notes that behavioral ecologists often have tested multiple alternative hypotheses, and suggests ways for them to do so more often and more effectively.

So there's one reason to think that ecologists are missing some opportunities to test alternative scientific hypotheses: we've often done so before.

Another reason to think that ecologists have room to improve when it comes to testing alternative scientific hypotheses is the apparent lack of progress in settling some long-standing ecological controversies. Indeed, controversies in our field about which alternative scientific hypothesis is correct sometimes get *more* controversial the more we learn about them.

Back in 2018, I polled *Dynamic Ecology* readers about their views on 22 controversial topics in ecology (J. Fox 2018). Importantly, all the topics were of long-standing interest in ecology. They've all been the subject of decades of research. For each topic, I provided a one-sentence summary statement of one leading hypothesis or other claim—for instance, "Habitat fragmentation per se, as opposed to habitat loss, typically reduces biodiversity." I asked respondents to indicate their agreement or disagreement with that statement on a scale of 1 ("definitely false") to 5 ("definitely true"). I also asked respondents to indicate their own level of expertise on each topic on a scale from 1 to 4: "1: I know nothing," "2: I know a bit (e.g., learned about it in a class)," "3: I know some," and "4: I'm an expert." The poll got 251 respondents. Some topics turned out to be quite controversial, a few weren't controversial at all, and many were somewhere in the middle. But the most relevant, and striking, result for present purposes was that almost every topic was more controversial among *experts* on that topic than among ecologists as a whole. That is, the experts—the people who know the most about the topic—almost invariably disagreed the most as to the truth of the statements (fig. 4.1). Further, it's not just that the experts disagreed with one another more than the non-experts. Rather, the experts sometimes disagreed massively in absolute terms. They sometimes split into opposing camps, one of which was pretty confident the statement was true and the other of which was pretty confident it was false (fig. 4.2). And remember, all these topics are heavily researched. So it's not that the experts disagreed because they're the ones who are most acutely aware of the limitations of the available evidence. There's more evidence about these topics than there is about the typical ecological topic! Note as well that it's not that the experts (or non-experts) all agreed that the claims on which I polled are true in some contexts but false in others. Nor did they all agree that the available data are too limited and messy to draw any firm conclusions. That is, they didn't all choose 3 on the 1–5 scale I provided.

Now, these poll results provide only a snapshot of (some) ecologists' opinions, rather than documenting changes in opinion over time as more data comes in. Still, I think it's striking—and sobering—that some ecological topics can remain so controversial among experts after decades of research. If we spend decades studying a topic only to end up with opposing camps favoring opposing hypotheses, something has gone wrong somewhere along the line. Opposing scientific claims provide fuel for selection, but for some reason, the selection often doesn't happen. It's as if the relative abundances of different plant species in a diverse plot somehow became entrenched over time, resistant to any further change.

Figure 4.1. Experts on controversial ecological topics exhibit greater variance of opinion about those topics than do non-experts

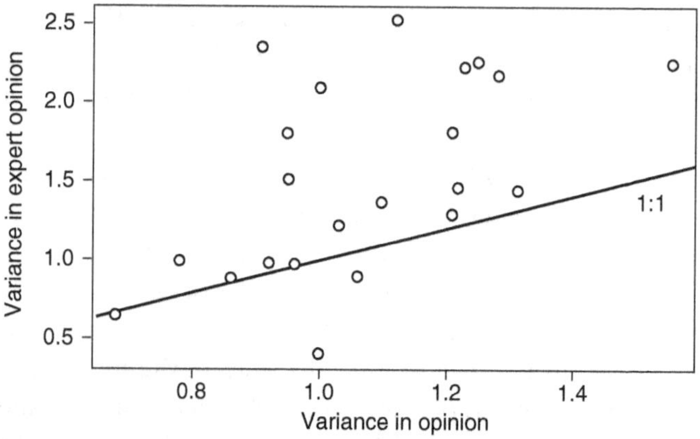

The graph plots variance of opinion among experts versus the variance of opinion among all ecologists. Each point gives data for one controversial topic. The solid line is the 1:1 line. There is more variance in opinion among experts than among all ecologists (experts and non-experts) for 19 out of 22 topics.

Source: J. Fox (2018).

Making Selection Work in a Messy, Complicated World: Journals Selecting Which Papers to Publish

Selecting from among alternative scientific hypotheses in ecology isn't the only context in which selection is challenging (though perhaps not as challenging as some have claimed). Let's turn our attention from selection in the scientific research process to selection in the publication process. Peer review is the process scientists use to select which papers to publish in which journals. It's a challenging task. But peer review in ecology actually works pretty well, despite facing many of the same challenges that make it difficult to select from among alternative scientific hypotheses in ecology.

You don't have to hang around scientists very long before you hear complaints about the peer-review system (R. Smith 2006; Belluz 2016). Many of those complaints concern peer review selecting for the wrong things. There are complaints that it selects for novel results over replications of previous work (H. Fraser, Barnett, et al. 2020). There are complaints that it selects for flashy but unsound papers over boring but

Figure 4.2. Histograms of ecologists' opinions regarding two controversial research topics

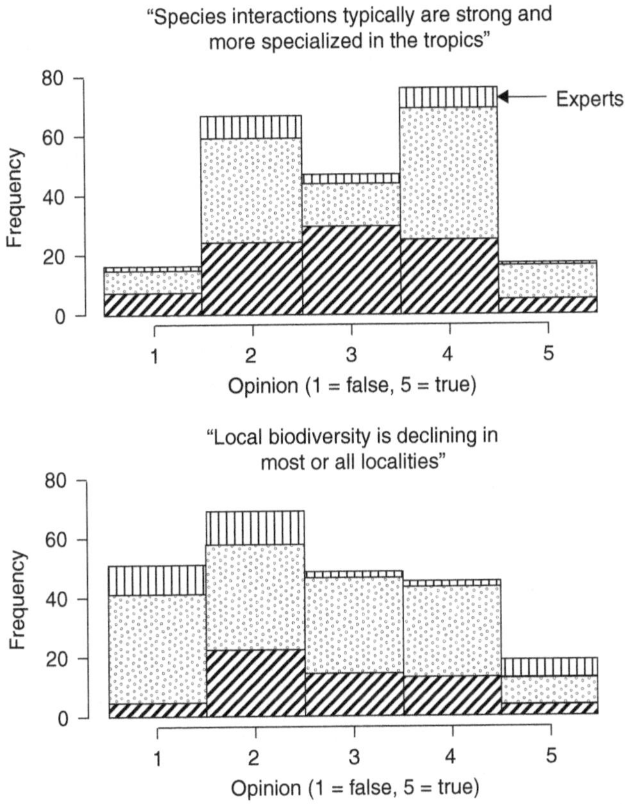

Opinions of experts (*vertical stripes*), those who know something about the topic (*stippled*), and those who know a bit about the topic (*diagonal stripes*). Expert opinion on both topics is bimodal, and expert opinion is more bimodal than non-expert opinion. This indicates substantial disagreement among experts, exceeding that among non-experts.

Source: J. Fox (2018).

rigorous papers (Belluz 2016). There are complaints that it selects for papers by famous authors (Huber et al. 2022). And so on.

But another common complaint about peer review is the opposite: that it isn't selective at all (Rothwell and Martyn 2000; D. Peters and Ceci 1982). Rather, peer review is a lottery. Just a matter of luck. Whether your paper gets accepted or rejected might depend on whether the reviewers assigned to your paper are lazy. It might depend on whether you're unlucky enough to be assigned an unreasonably picky reviewer. It might depend on whether you're unlucky enough to be assigned a reviewer from a different subfield than your own, who just doesn't understand the point of your paper. It might depend on the editor's highly subjective (and thus, effectively random) judgment about whether your paper is "interesting." And so on. On the "lottery" view, the measurable outcomes of peer review should be unpredictable, just as the winner of a lottery is unpredictable.

So which is it, in ecology? Is peer review in ecology selective, or a lottery, or some mix of the two? If it's a mix of the two (which seems plausible), is it more of a lottery at some journals than others? And to the extent that peer review *is* selective, *what* does it select for?

We're fortunate to have excellent data addressing these questions for ecology journals. Charles Fox (no relation), former executive editor of *Functional Ecology*, has used data from ecology journals' online manuscript-handling systems, large randomized surveys of authors, and large controlled randomized experiments to address numerous questions about peer review in ecology (C. Fox, Paine, and Sauterey 2016; C. Fox 2021, 2017; C. Fox and Paine 2019; C. Fox, Meyer, and Aimé 2023; C. Fox, Duffy, et al. 2019; C. Fox, Burns, et al. 2017; C. Paine and Fox 2018; Burns and Fox 2017). You should read all his papers, along with the follow-up analyses others have done using the same data (e.g., Srivastava et al. 2024) and similar analyses of other datasets (Campos-Arceiz, Primack, and Koh 2015). Here, I'll just focus on one set of results that quantify the extent to which peer review in ecology is a lottery.

Tim Paine and Charles Fox (2018) created a stratified random sample of tens of thousands of ecology papers published in 146 ecology journals from 2009 to 2015. They then surveyed the corresponding authors of those papers (one paper per corresponding author) regarding the paper's peer-review history. Was the paper published in the first journal to which it was submitted? If not, at which journal(s) was it previously rejected, and when? If it was rejected previously, was it rejected before peer review ("desk rejection") or after? After filtering out nonresponses, incomplete responses, invited papers, and review papers, Paine and Fox ended up with data on 10,580 papers by the same number of unique corresponding authors. Those papers had gone through a total of 16,981 rounds of manuscript submission. Paine and Fox also compiled data on how often those 10,580 papers were cited, and on how often the typical paper published in a given year by a given journal was cited. These data allowed the authors to ask if, and how, a paper's peer-review history is associated with how often the paper is cited, compared with a typical paper published in the same journal in the same year.

If peer review is a pure lottery, then there shouldn't be any relationship between a paper's peer-review history and where it's eventually published or the number of times it's cited. But in fact, there are some pretty strong relationships. Peer-review outcomes aren't perfectly predictable, so peer review in ecology *is* something of a lottery. But it's far from being a pure lottery. Overall, peer review in ecology is pretty good at selecting for manuscripts that will be highly cited, and against manuscripts that will not be.

Most rejected papers are eventually published in journals with lower impact factor (IF) than the journal to which they were originally submitted (fig. 4.3). Further, the comparatively rare rejected papers that are eventually published in a journal of similar or higher impact to the original (rejecting) journal obtain many fewer citations on average than the typical paper published in the same year and journal (fig. 4.3). Averaging over all journals, rejected papers garnered only 40% as many citations as the typical paper published in the same journal in the same year. Finally, only about 9% of rejected papers went on to become high-impact papers. These results indicate that rejected papers aren't rejected at random. Rather, reviewers and editors who select papers for rejection do so on the basis of real, identifiable features of the papers they reject. Presumably, those same features, or other tightly correlated features, explain why those papers go on to garner relatively few citations, even if they do manage to get published in a high-impact journal.

You also can turn the analysis around: ask how well the number of citations a paper eventually garners predicts its peer-review history. High-impact papers were only 23% as likely to have been desk-rejected as run of the mill papers, and were only 41% as likely to have been rejected after peer review.

Finally, the highest-impact journals have the highest rejection rates. They're also the best at distinguishing high-impact papers from low-impact papers, by various measures. That is, to the extent that peer review is something of a lottery, it's actually less of a lottery at journals that

Figure 4.3. Manuscripts rejected from one ecology journal and resubmitted to another obtain fewer citations on average than does the typical paper accepted by the journal to which it was originally submitted (performance ratio <1)

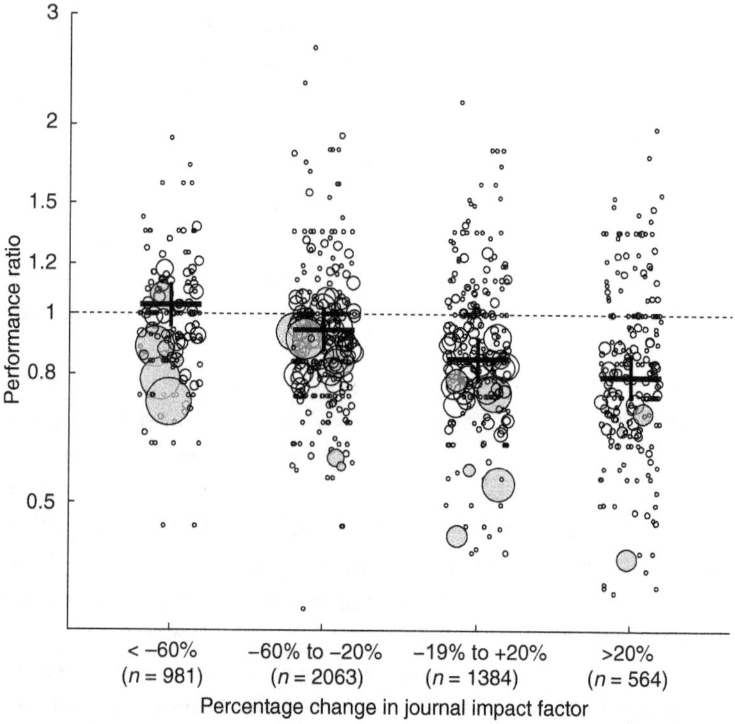

Performance ratio also depends on the impact factor of the publishing journal relative to that of the journal that originally rejected the paper (*x*-axis). Rejected manuscripts resubmitted to journals with impact factor >20% higher than that of the original journal are rare and tend to exhibit especially low performance ratios (*x*-axis, *rightmost column of points*). Crosses indicate means (*horizontal lines*) and bootstrapped 95% confidence intervals (*vertical lines*). All groups differ significantly from all others in mean performance ratio.

Source: Modified from C. Paine and Fox (2018).

reject a larger fraction of submissions. It may be less of a lottery at high-impact journals, because high-impact journals have more variation to select among. Many authors have professional incentives to aim high when submitting manuscripts to journals, even at the cost of submitting to a journal that's a poor fit. Authors also may tend to overrate the quality of their own papers (Rastogi et al. 2024). Both of these factors would cause the highest-impact journals to receive the widest range of manuscripts—those destined to become low-impact papers as well as those destined to become high-impact ones. As C. Paine and Fox (2018, 9574) put it, "Our results suggest that researchers often submit their manuscripts to journals for which they are unsuitable, but that editors and reviewers are generally good at identifying and rejecting such papers."

Taken at face value, all that the results of Paine and Fox show is that peer reviewers, editors, and citing authors all select for the *same* features of ecology papers. Their results don't speak to what those features are or whether they're good features to select for. My own view, based on my own experiences as an author, peer reviewer, and editor, is that peer reviewers and editors primarily select for the features that the journals *say* that they select for, in their instructions for authors. At high-impact journals, those are features like generality, novelty, and importance. And I think it's good to select for those things! However, like the authors of H. Fraser, Barnett, et al. (2020), I also think there's room to argue that ecology journals ought to select more heavily for replicability and robustness.

I take heart from the fact that peer review in ecology is a fairly effective example of selection. It selects pretty successfully for the things it's intended to select for. Is it perfect? No. Can it be improved? Sure—improvements get proposed and implemented all the time. For instance, many ecology journals implemented double-blind review within my professional lifetime. But is peer review anywhere close to being a pure lottery? No.

Peer review reminds us that it's perfectly possible to select from among alternatives, even when they differ from one another in all sorts of ways, have features that are difficult or impossible to quantify, and aren't mutually exclusive. I think that's a useful reminder. The most common excuses for why Plattian strong inference is rarely possible in ecology—ecological systems are messy and complicated, alternative hypotheses aren't mutually exclusive, decisive experiments are infeasible and so on—prove too much. Because those excuses apply just as much to peer review of scientific papers! Indeed, they apply *so* well to the peer-review system that they make its very existence quite surprising. If the peer-review system didn't exist, we would all think it was impossible, *and with good reason*. How could you possibly design a system that does even a halfway decent job of selecting from among millions of scientific papers every year, on the basis of various fuzzy, hard-to-measure features like "generality," "novelty," "importance," and "technical soundness"? Especially a system based on fallible, busy, biased humans, rather than, say, a computer algorithm? Even if you could somehow design and organize the whole system and get millions of scientists to all buy in, surely the results would be no better than a lottery!

So don't be too quick to buy into even extremely reasonable objections to the possibility of testing alternative hypotheses in ecology. Some things seem impossible because they are in fact impossible. But others seem impossible right up until they turn out not to be.

Selecting for the Wrong Things 1: Authors Selecting Which Results to Publish

In the second half of this chapter, let's turn our attention from failure to select, to selecting for the wrong things—analogous to a negative selection effect in a BEF experiment in which initially diverse plots come to be dominated by the least-productive species.

What gets published in scientific journals depends on the selective choices of authors, not just peer reviewers and editors. Like most ecological researchers, I haven't published every study I've ever done or supervised. I haven't even tried to do so. Some I'll get around to publishing eventually. Case in point: recently I published results of an experiment that my lab did more than a decade earlier (J. Fox 2023). But other studies I'll never publish at all, mostly because the results have a low signal-to-noise ratio. They aren't distinguishable from sampling error. I can't be bothered to publish such results, because I'm limited by time, money, and interest. From my perspective, there'd be too much opportunity cost to spend many hours, and hundreds or thousands of dollars in article-processing charges, to publish a boring, nonsignificant result.

That's the best choice for me, or at least a defensible choice. I'm far from alone in thinking this way (e.g., Duffy 2013). But the choice not to publish some nonsignificant results has costs for the field of ecology as a whole. It leads to the file drawer problem: published studies are a statistically biased subset of all studies. If studies that fail to find a significant effect of *x* tend to go unpublished, then the published literature will overstate the frequency with which effect *x* is statistically significant, and overstate the typical magnitude of effect *x*. The file drawer problem is a form of selection effect. Authors select results to publish, and they make that selection nonrandomly with respect to features of the results, such as the magnitude of the effect or its statistical significance.

How big a problem is the file drawer problem in ecology? Until fairly recently, it was hard to say. Julia Koricheva (2003) examined a sample of Swedish and Finnish PhD theses in ecology, produced from 1982 to 1988. She asked if thesis chapters reporting a greater proportion of nonsignificant results were more likely to remain unpublished in scientific journals. She also surveyed the authors to ask them why their unpublished chapters remained unpublished. The frequency of nonsignificant results didn't vary between published and unpublished thesis chapters. Lack of statistically significant results was a rare reason for authors not to submit chapters to journals, and for journals to reject chapters for publication. Koricheva's results suggest that the file drawer problem isn't much of a problem in ecology, but it's not clear if those results generalize. And the results are admirably information rich—unfortunately, at the cost of a modest sample size. It's hard to scale up such a labor-intensive research approach to the entire field of ecology.

An alternative approach to quantifying the file drawer problem *can* be scaled up. It's a statistical approach known as Egger's regression (Egger et al. 1997). Egger's regression is a way to test for publication bias in published studies on the same topic, such as the studies included in a meta-analysis. We can summarize each study with two numbers: an "effect size" and the standard error of that effect size. An effect size is a measure of the study outcome, such as the log-transformed ratio of treatment-group mean to control-group mean. Its standard error is a measure of precision. If you repeat the same study on the same statistical population again, how much would you expect the effect size to vary among repetitions? In the absence of publication bias, plotting the effect sizes of different studies versus their standard errors should give you a funnel-shaped plot (fig. 4.4, upper left). That's why it's called a "funnel plot." Studies with small sample sizes produce widely varying, imprecise estimates of the population mean effect size. Studies with large sample sizes produce precise effect size estimates, all clustered close to the true population mean. If we were to conduct a linear regression of effect sizes (appropriately scaled) on the inverse of their standard errors, the intercept of the regression would be close to zero (fig. 4.4, upper right). Conversely, nonzero intercepts are a statistical symptom of an asymmetrical funnel plot, and an asymmetrical funnel plot suggests (although doesn't prove) a file drawer problem (fig. 4.4, bottom panels).

This simple picture gets more complicated if there's heterogeneity—that is, the true mean effect size varies among studies. Heterogeneity makes it more difficult to detect publication bias. You don't necessarily expect a funnel plot to look funnel-shaped in the presence of strong heterogeneity. But I'm going to set that complication aside because, as we'll see, it turns out that publication bias in ecology is still detectable against a background of heterogeneity.

Figure 4.4 shows simulated examples of publication bias. You can find similarly strong examples in the real world if you look for them.

Figure 4.4. Illustrations of publication bias and the lack thereof

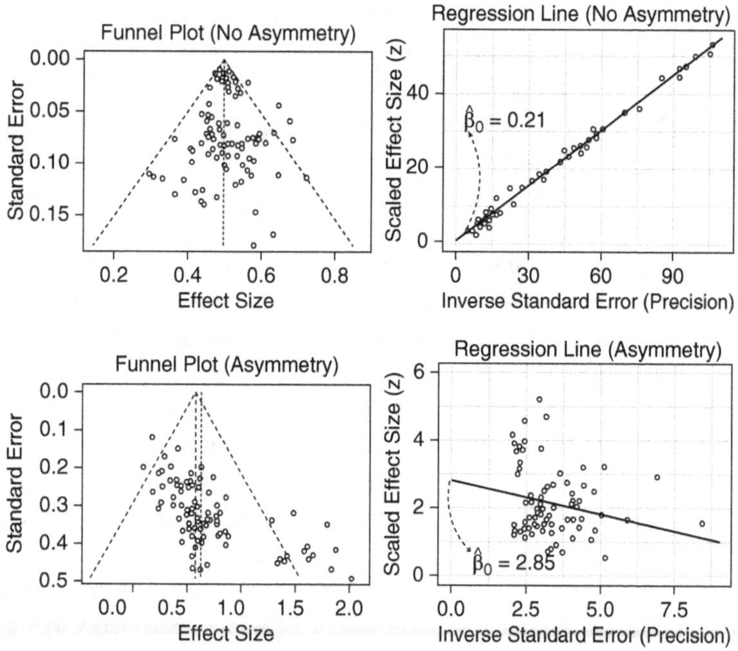

Funnel plots (*left column*), and Egger's regressions (*right column*), illustrating lack of publication bias (*top row*), and publication bias against nonsignificant effects (*bottom row*). Nonsignificant effects mostly come from small studies.

Source: Harrer et al. (2021).

Figure 4.5 shows some examples from the literature on the ecological effects of microplastics. Hundreds of published observational and experimental studies show that ingestion of microscopic plastic particles has negative effects on organismal behavior, growth, survival, and reproduction. But many small studies finding neutral or even positive effects of microplastics apparently have gone unpublished.

Examples of strong publication bias are good examples for pedagogical purposes. They're easier to grasp than less clear-cut examples would be. That doesn't necessarily make them typical examples, though. So how common and strong is publication bias in ecology? How big and full is our proverbial file drawer?

Yang, Sánchez-Tójar, et al. (2023) addresses that question using data from 87 ecological and evolutionary meta-analyses covering a wide range of topics. Fifty of those meta-analyses originally reported a mean effect size significantly different from zero. Using a modified version of Egger's regression, the authors found that 17% of meta-analyses exhibited statistically significant publication bias against small studies reporting nonsignificant effects—well above the 5% that one would expect if no meta-analysis truly exhibited any publication bias. The distribution of estimated publication-bias effects (statistically significant or otherwise) was skewed toward publication bias against small studies reporting nonsignificant effects, and the grand mean was significantly different from zero (fig. 4.6). The absolute magnitude of the estimated mean effect size often dropped by more than 50% after correcting for estimated publication bias, and 33/50 statistically significant mean effect sizes became nonsignificant after correction.

So, yes, the file drawer problem often is big enough to alter the most basic inferences one can draw from the ecological literature, such as "Is this effect nonzero, on average?" or "Roughly how strong is this effect on average, within a factor of two?" Empirical ecological research produces a diverse range of results on any given topic. And then ecologists—including me!—select a nonrandom subset of those diverse results to publish.

What can be done about this? One option is to make it easier, faster, cheaper, and more rewarding to publish nonsignificant results. The advent of author-pays open-access journals that publish any technically sound paper certainly has made it easier and faster to publish

Figure 4.5. Publication bias in studies of the effects of microplastics on organismal behavior, growth, reproduction, and survival, as shown by asymmetrical funnel plots

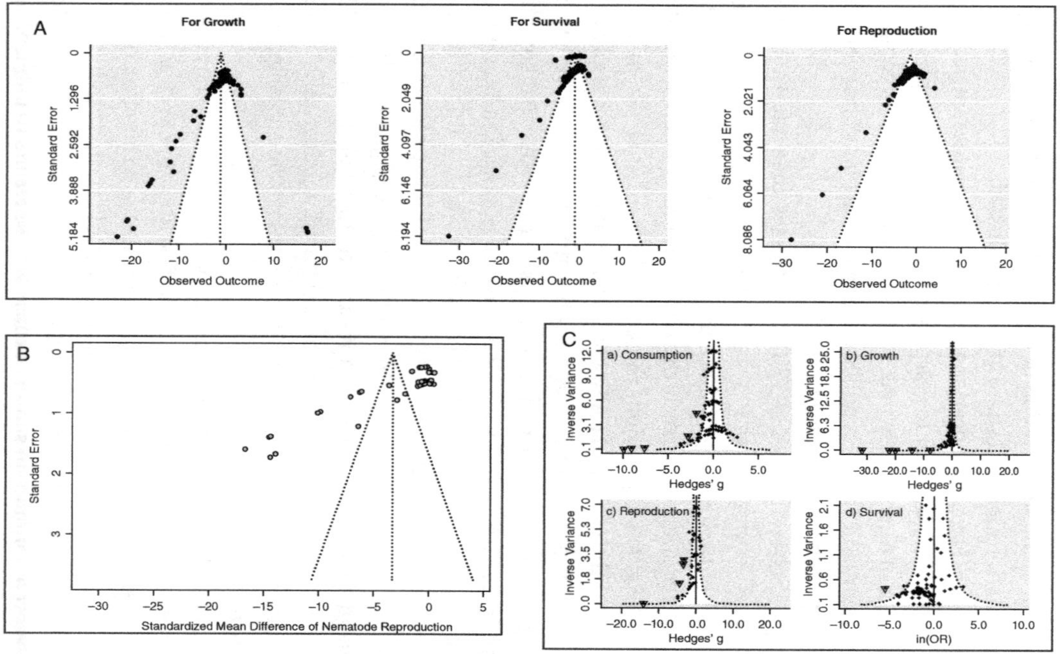

The panels are modified from meta-analyses of the effects of microplastics on (A) soil fauna, (B) soil invertebrates, and (C) fish and aquatic invertebrates.

Source: Modified from (A) Wan et al. (2023), (B) Ji et al. (2021), and (C) Foley et al. (2018).

Figure 4.6. Ecological and evolutionary meta-analyses tend to exhibit publication bias against small studies that report nonsignificant effects

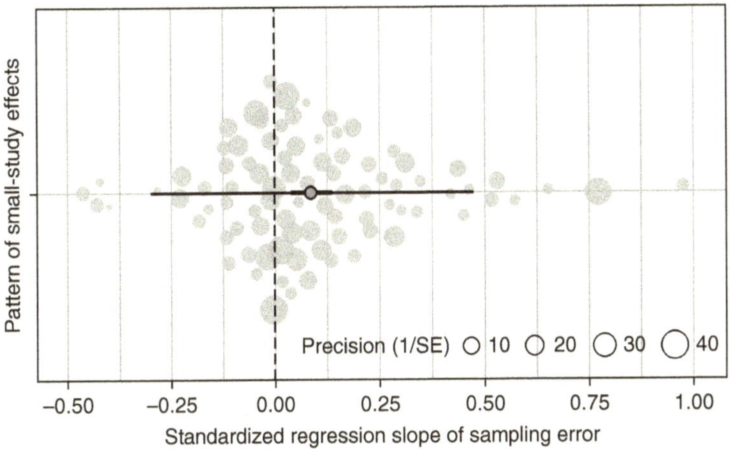

Orchard plot of the distribution of "small-study effects" (slopes of standardized regressions of sampling error vs. effect size) in 87 ecological and evolutionary meta-analyses. Positive regression slopes are consistent with publication bias against small studies that report nonsignificant effects. Slope estimates come from a larger mixed model that allows for heterogeneity and non-independence of effect sizes within and among primary research papers.

Source: Modified from Yang, Sánchez-Tójar, et al. (2023).

nonsignificant results, though "cheaper" and "more rewarding" remain works in progress (McGill 2024). But in the absence of a fix, I think ecologists should just routinely assume that the ecological literature on any topic is distorted by publication bias, and mentally adjust the results of published meta-analyses accordingly. For instance, the results of Yang, Sánchez-Tójar, et al. (2023) suggest that a good rough-and-ready rule of thumb for ecologists is "The true mean effect size is only half as large as the meta-analytic estimate."

Selecting for the Wrong Things 2: Authors Selecting Which Results to Highlight

Selecting among diverse results doesn't stop once you've selected which results to publish. Next, you have to select which results to highlight as your main results. Which results will you emphasize in your paper title and abstract? Which results will you summarize in less detail in the abstract? Which results will you mention only in passing in the main text of the paper, without any mention in the title or abstract? And which results will you bury in online appendixes?

Those choices matter because it's not just authors who are selective—readers are too. The typical ecology paper isn't likely to receive a careful, line-by-line read from anyone besides a couple of pre-publication peer reviewers. Most readers are going to read the title and abstract, glance at the figures, and *maybe* skim the Results section. Or else they won't even do that; they'll just see a brief summary of the take-home message on social media. None of which is a criticism of readers, by the way! Reading an entire scientific paper carefully often isn't a good use of a reader's time. Part of the point of the pre-publication peer-review system is to distribute the work of careful reading, thereby reducing the need for *everyone* who reads a given paper to do so carefully. So, as an author, your choices as to which results to highlight in your title and abstract go a long way toward determining what message readers will take from your paper.

For a long time, I thought authorial choices as to which results to highlight were some mixture of straightforward and personal. Straightforward: for many papers, the choice of which results to highlight is obvious; any competent author would make the same choice. Personal: in cases where the choice of which results to highlight isn't obvious, different authors will have different preferences. Those preferences will be harmlessly personal and idiosyncratic, like preferences for different flavors of ice cream. But recently I've started wondering whether there isn't a third possibility: cases where the choice of which results to highlight isn't obvious yet most authors make similar choices—with the consequence that readers who aren't reading carefully (which is most of them) end up with a skewed, inappropriately selective picture of published results.

I started wondering about this thanks to the work of Lenore Fahrig on habitat fragmentation and biodiversity. Habitat fragmentation refers to breaking up contiguous habitat into smaller patches, separated by boundaries, such as roads or some other distinct type of habitat ("matrix" habitat). Habitat fragmentation is often associated with—but is conceptually distinct from—habitat loss. Reduction of total habitat area and breaking the same amount of habitat area into smaller patches are two different things, even if they often co-occur. How does habitat fragmentation per se affect biodiversity? That is, given two landscapes with the same total area of habitat, and that are similar in other respects, will the more- or less-fragmented landscape support higher biodiversity? This question is of applied interest because of its relevance to the SLOSS question, the choice between "*s*ingle *l*arge *o*r *s*everal *s*mall" nature reserves (reviewed in Fahrig et al. 2022).

In theory, one could imagine the effect of fragmentation going either way. For instance, some species prefer habitat edges, so fragmentation might increase biodiversity by increasing the species richness and abundance of edge-associated species (Willmer, Püttker, and Prevedello 2022). Fragmented habitats might allow for metapopulation dynamics that allow persistence of some species that would otherwise go extinct in unfragmented habitat (Yaari et al. 2012). Conversely, some large, highly mobile, edge-averse animal species might need large areas of unfragmented habitat to persist (Runge et al. 2014). Populations within small habitat patches might be vulnerable to extinction via inbreeding and demographic stochasticity, while the inhospitable interpatch matrix prevents recolonization from other patches (Shafer 1995). The effects of fragmentation might also depend on spatial scale. For instance, fragmentation might reduce the mean number of species within a typical patch but increase the number of species supported by the entire landscape, because different patches support different species. All those possibilities and more could be operating at once. Different possibilities might predominate for different species, or for different habitat types. And so on. So you might expect the empirical literature on effects of habitat fragmentation on biodiversity to report mixed results.

And indeed, it does—but you have to read very closely to learn that, rather than just reading the results that authors select for highlighting in their abstracts. Lenore Fahrig (2017a, 2017b) identified 118 studies (primary research papers) of the effect of habitat fragmentation per se, as distinct from habitat loss. Those studies reported a total of 381 statistically significant effects of fragmentation on some measure of biodiversity, 76% of which were positive. Only nine studies found exclusively negative effects of fragmentation on biodiversity among their statistically significant results, versus 84 studies for which all the statistically significant effects were positive. Further, positive effects of fragmentation were disproportionately likely for the species of greatest conservation concern: habitat specialists and threatened species. So, mixed results, with positive effects clearly predominating. But read the abstracts of those 118 studies and you'll get a rather different message. Seven of nine studies that found exclusively negative statistically significant effects of fragmentation highlighted those negative effects in their abstracts. In contrast, only 34 of 84 studies that found exclusively positive statistically significant effects of fragmentation described their own results as positive in their abstracts. Instead, the majority of those 84 studies either didn't mention the positive effects at all, or else presented them as negative or mixed. Finally, the minority of studies that found exclusively positive statistically significant effects, and that mentioned those positive effects in their abstracts, often emphasized caveats in their abstracts—for instance, emphasizing that the positive effects of fragmentation might be atypical, or might not generalize to other studies. In contrast, none of the studies that found exclusively negative statistically significant effects mentioned any caveats in their abstracts.

The results of Fahrig (2017a, 2017b) are consistent with other lines of evidence on the selectivity of authors of habitat fragmentation papers. For instance, Federico Riva, Nicola Koper, and Fahrig (2024) used automated text mining and sentiment analysis to show that effects of habitat fragmentation are discussed in as negative a tone as, say, resource overexploitation or invasive species. Further, papers that discuss habitat fragmentation in a more negative tone tend to be cited more often, even after controlling for the impact factors of the journals in which they were published, their geographic scope, and their taxonomic scope (Riva, Koper, and Fahrig 2024). Ecologists studying the effects of habitat fragmentation on biodiversity obtain a diverse range of mostly positive effects. Then they systematically select the negative effects for emphasis and are rewarded for their selectivity with citations.

None of which means that habitat fragmentation researchers are bad scientists who are consciously aiming to mislead their readers! Rather, they're just doing two things most ecological researchers (including me) do all the time: (i) interpreting mixed results in line with their preexisting expectations, and (ii) trying to make sure their results aren't misinterpreted or misused by managers or policymakers.

The trouble with (i) isn't just these ecologists' reluctance to give up on their preexisting expectations in light of contrary data; it's also the existence of those preexisting expectations in the first place. Ecologists often write papers as if we have strong preexisting expectations about the results we're likely to obtain—for instance, writing as if we have strong reason to expect the effects of habitat fragmentation on biodiversity to be negative, even though there are theoretical reasons to expect positive as well as negative effects. It's arguably good scientific practice to discount "unusual" or "weird" results that conflict with one's preexisting expectations—so long as those preexisting expectations are well justified. There's a good reason why physicists who found evidence of neutrinos traveling faster than the speed of light refused to believe their own result, even before they discovered the reason for that result (Kim 2024). But are ecologists' preexisting expectations—about the effects of habitat fragmentation on biodiversity, or about the effect of anything on anything else—well justified? As noted above, there are theoretical reasons to expect the effects of habitat fragmentation on biodiversity to cut both ways. More broadly, literally *every* empirical research topic ecologists have studied in more than a hundred papers has yielded quite variable heterogeneous results, even for matters as basic as the sign of the effect of one variable on another (see chapter 7). There's almost literally no such thing as an "unusual" or "weird" empirical result in ecology, whether we're talking about the effect of habitat fragmentation on biodiversity, or the effects of anything on anything else. So shouldn't our preexisting expectation about the effect of *x* on *y* always be ¯_(ツ)_/¯?

The trouble with (ii) is that, in practice, "making sure my results aren't misinterpreted or misused" tends to end up meaning "presenting my results so that they align with the policy preferences of ecologists." And one respect in which ecologists are *not* especially diverse, at least compared to the general public, is with respect to their preferences regarding environmental policies. For instance, habitat fragmentation researchers who find positive effects of habitat fragmentation on biodiversity likely want to ensure that their results aren't misinterpreted or misused to, say, justify road building. So the researchers selectively emphasize their results in order to prevent such misinterpretation and misuse. Thereby unintentionally promoting a different, equally bad misinterpretation: the idea that small, isolated habitat fragments have no conservation value (Fahrig 2017a, 2017b).

Just speaking personally, these data on how ecologists selectively present the results of habitat fragmentation research are some of the most important—and worrisome—data I've ever seen in ecology. An important task for future work is to check whether these results generalize to other areas of ecological research. My guess (and that's all it is) is that they do generalize, at least to some extent. It just seems unlikely to me that Fahrig stumbled across the only example of systemic selective presentation of results in all of ecology. But I hope I'm wrong about that.

These results highlight the value of diversity of ideas in ecology, even in—perhaps especially in—the face of a strong fieldwide consensus. Scientific research is supposed to be self-correcting, so that it converges on the truth in the long run. But some mechanisms of self-correction break down when researchers approach, or achieve, consensus. That's a good thing, when the consensus matches the truth. But it's a bad thing when the scientific consensus has significant room for improvement. As we'll see in chapter 6, the marginal value of increased diversity of ideas is highest when the diversity of ideas is low. But it's when the diversity of ideas is low that it's hardest to increase it.

Selecting for the Wrong Things 3: Selecting Ineffective Methods for Inferring Process from Pattern

We've now seen two examples analogous to negative selection effects in BEF research: authors selecting which results to publish, and which results to emphasize within their publications, on the basis of understandable but questionable criteria. Let's consider one more example—really, one more set of examples. This last set involves negative selection effects in the context of research methods rather than of publishing completed research.

As we saw in chapter 1, ecologists sometimes want to explain *why* some bit of nature is the way it is. Why is there a latitudinal species richness gradient (Willig, Kaufman, and Stevens 2003)? How does the interplay of density-dependent and density-independent processes generate observed population dynamics (Bjørnstad and Grenfell 2001)? Does interspecific competition generate checkerboard distributions of competing species (Diamond 1975)? What mechanisms set the limits to species' geographic ranges (Hargreaves, Samis, and Eckert 2014)? All these questions ask about the processes that cause the patterns we observe in nature.

The best way to answer such questions is with numerous complementary lines of evidence, including but not limited to controlled, randomized, replicated experiments (Grace 2024; see also chapter 3). But we don't always live in the best of all possible worlds; Often, all we have to go on are the patterns themselves. We can't run any controlled, randomized, replicated experiments, for whatever reason. What if we still want to infer the processes that caused the patterns we observe?

Ecologists who want to infer process from pattern, as opposed to doing experiments to help infer causality, often cite examples of such inferences from other fields of science. James Brown (1995, 21) reported that Robert MacArthur once said to him, "Astronomy was a respected, rigorous science long before ecology was, but Copernicus and Galileo never moved a star." Brown cited plate tectonics as a second example of a successful inference of process from pattern. Brian Maurer (1999) also cited astronomy as an example of a science that routinely infers process from pattern without resort to experimental manipulations of its primary objects of study (stars, galaxies, etc.)

The examples of astronomy and plate tectonics do indeed establish that experiments are not the only way to do rigorous, process-inferring science (which is probably all they were intended to establish). But they don't establish that ecologists can infer process from pattern, much less that ecologists can do so using the approaches taken by astronomers and geologists.

In this section I consider successful and unsuccessful attempts to infer process from pattern, from ecology and other fields of science. Ecologists, as well as scientists in other fields, have used a diverse range of research approaches to successfully infer process from pattern. But too often, ecologists haven't selected the approaches that have worked in the past. Instead, we too often have selected approaches that have repeatedly failed to work in the past. The upshot is that when it comes to selecting methods for inferring process from pattern, the negative selection effects generated by ecologists' methodological choices aren't idiosyncratic. Rather, they tend to run to type. Hopefully, recognizing that these negative selection effects run to type can help us avoid them in the future.

METHOD FOR INFERRING PROCESS FROM PATTERN 1: MEDICAL DIAGNOSIS

Medical diagnosis is one of the most familiar examples of inferring process from pattern. Your doctor asks you about your symptoms and your medical history. Those observations make up the observed "pattern." On the basis of the observed pattern, plus relevant background information (e.g., knowledge that some diseases are much more common than others), the doctor diagnoses the causes of your symptoms, or at least narrows down the potential causes (e.g., fig. 4.7).

Medical diagnosis works only when the combination of symptoms and background information is, well, diagnostic: there's a one-to-one mapping between symptoms and cause, pattern and process. That is, each unique combination of symptoms plus background information corresponds to a unique affliction. Medical diagnosis doesn't work if the observed symptoms and background information are consistent with many different afflictions. That's why doctors often use diagnoses

Figure 4.7. Common symptoms of infection with the cold virus, influenza virus, and COVID-19 virus

Common Symptoms of a Cold, the Flu, and COVID-19
Learn more at www.nia.nih.gov/flu
NIH National Institute on Aging
*Symptoms may vary based on new COVID-19 variants and vaccination status.

Common Symptoms*	Cold	Flu	COVID-19
Fever and/or chills		✓	✓
Headache		✓	✓
Muscle pain or body aches		✓	✓
Feeling tired or weak		✓	✓
Sore throat	✓	✓	✓
Runny or stuffy nose	✓	✓	✓
Sneezing	✓		
Cough	✓	✓	✓
Shortness of breath or difficulty breathing		✓	✓
Vomiting and diarrhea		✓	✓
Change in or loss of taste or smell			✓

Source: US National Institutes of Health.

based on symptoms, medical history, and other relevant background information only as a starting point. They often follow up a tentative or partial diagnosis with diagnostic tests to further narrow down the causes of the patient's symptoms.

Ecologists often attempt to do the same thing: infer the process(es) that caused the patterns in the observed data by ruling out other processes that aren't consistent with the observed patterns and our background knowledge. Population ecologist that I am, one of my favorite examples is to do with inferring the causes of population cycles.

Ives, Einarsson, et al. (2008) provides an example of "medical diagnosis" of the causes of population cycles. Larvae of the midge *Tanytarsus gracilentus* are the dominant herbivores and detritivores in Lake Myvatn in Iceland. Adult midges emerge from the water to breed near the lake. Adult midge abundances exhibit high-amplitude fluctuations (six orders of magnitude), but with an irregular period of four to seven years. Those are the "symptoms" the authors of Ives, Einarsson, et al. (2008) set out to "diagnose." The authors fit three alternative models to the observed midge dynamics. All three models described the interactions between midge larvae and their food resources (algae and detritus), but differed in the details of how they did so. Crucially, only one of the three models was capable of generating what are known as "alternate attractors." For realistic parameter values, this model predicted that midge abundances in the future could remain consistently high, or else exhibit high-amplitude cycles with a regular period. Future midge dynamics could follow either of those two "attractors," depending on the current abundances of midges and their food resources. Environmental stochasticity—random fluctuations in the weather and other abiotic variables that affect midge birth and death rates—would then cause midge abundances to fluctuate irregularly by switching the system between those two attractors. When midges are at high abundance, the population is balanced on a figurative knife edge. Depending on which way the abundance "wobbles" due to environmental stochasticity, it can either remain at high abundance in the near future or start cycling between high and low abundances. The alternate attractor model proved much better than the other two models at predicting midge abundance in generation $t + 1$ after being fit to the first t generations of data. That's despite the fact that the other two models were designed to be more flexible than the alternate attractor model. That is, the other two models were capable of producing a wider variety of population dynamics, not just high stable midge abundances or else a high-amplitude cycle. The alternate attractor model also ceased to fit the data well when fit under the constraint that the parameter values not generate alternate attractors. This provides additional evidence that the combination of alternate attractors and environmental stochasticity explains the irregular, high-amplitude fluctuations of this midge population.

But not all attempts by ecologists to do the equivalent of medical diagnosis are as effective as that of Ives, Einarsson, et al. (2008). Sometimes we try to do the equivalent of medical diagnosis without considering all relevant symptoms, or with a mistaken understanding of which afflictions cause which symptoms.

For instance, G. Evelyn Hutchinson and Robert MacArthur (1959) observed that if you measure the sizes of the feeding structures of sympatric congeneric species (e.g., bird beak sizes), you'll find that each species has a feeding structure about 1.3 times as large as that of the next-smallest species. Hutchinson and MacArthur suggested that this approximately constant ratio might reflect interspecific competition preventing the coexistence of species that are too similar in body size, and so could be taken as evidence for interspecific competition.

This suggestion inspired many ecologists to measure "Hutchinson's ratio" for many assemblages of putatively competing animals, and inspired models of "limiting similarity" of coexisting species (e.g., Hutchinson and MacArthur 1959; MacArthur and Levins 1967). Most empirical studies found size ratios around 1.3 and inferred interspecific competition as the explanation (Roth 1981; Gotelli and Graves 1996).

Unfortunately, Hutchinson's ratio doesn't allow you to infer anything about interspecific competition. The sizes of inanimate objects like bicycle wheels and frying pans also exhibit Hutchinson's ratios of 1.2 to 1.3 (Horn and May 1977; Maiorana 1978). Indeed, randomly sampling numbers from a lognormal or approximately lognormal distribution with a sufficiently small standard deviation (less than 1 or so) leads to adjacent numbers having ratios of approximately 1.3 (Eadie, Broekhoven, and Colgan 1987). The size distributions of species (and inanimate objects) studied by ecologists often fit the criteria of Eadie, Broekhoven, and Colgan (1987). In other words, saying that sympatric species exhibit

Hutchinson's ratios of around 1.3 is just a roundabout way of saying that sympatric species have roughly lognormal size distributions. You can't infer much of anything about the processes that generated a roughly lognormal distribution just from the fact that it's a roughly lognormal distribution. And you certainly can't infer interspecific competition.

Hutchinson's ratios are analogous to a medical diagnosis that fails to consider the full range of symptoms and all their possible causes. Finding a Hutchinson's ratio of ~1.3, and inferring interspecific competition from that finding, is like diagnosing a patient with COVID-19 just because the patient has a stuffy nose.

Here's another unsuccessful attempt to infer process from pattern using an approach analogous to medical diagnosis: local-regional richness relationships. The species richness of a local community (lake, area of grassland, etc.) varies from one community to the next. Typically, those species colonized that local site from somewhere else at some time in the past, usually from somewhere in the surrounding region. Of course, not all colonists necessarily succeed in establishing a local population. One reason they might fail is competition with resident species. So local species richness reflects both the influence of the surrounding region (the source of colonists) and local conditions that determine the fate of colonists. How can we tease apart the relative importance of local competition versus the supply of colonists from the regional "species pool" in determining local species richness?

Many ecologists have inferred the importance of competition in determining local species richness by compiling observational data on the richness of local communities, and the richness of the larger regions in which those communities are embedded, then regressing local species richness on regional richness (Terborgh and Faaborg 1980; Cornell 1985; Ricklefs 1987; Cornell and Lawton 1992). If the data are well described by a straight line with a positive slope, that's been thought to indicate that local communities are "samples" from the regional species pool. Each local community just contains some fraction of the species from the surrounding region, with the fraction being given by the slope of the regression line and reflecting the colonization rate. This implies that interspecific competition is too weak to limit local community membership. But if the data are better fit by a saturating curve (a curve that increases to an asymptote), that's been thought to indicate that local communities in sufficiently rich regions are "saturated" with species. The local "niche space" is full. Strong competition set an upper limit to local species richness that's independent of the number of species that could potentially colonize the local site. Linear and saturating local-regional richness relationships occur with approximately equal frequency in published studies (Gonçalves-Souza, Romero, and Cottenie 2013; Szava-Kovats, Ronk, and Pärtel 2013).

There are of course some potential technical problems with this approach. For instance, it's important that the regional species pool for a given local site not be defined so as to include species that couldn't possibly live at the site because they can't survive in the local abiotic conditions. And various confounding factors will affect the shape of the local-regional richness relationship (e.g., differences among regions in abiotic environmental conditions and evolutionary history; Srivastava 1999; Hillebrand 2005). Those are the normal sorts of surmountable technical problems any research approach in ecology faces.

But a more fundamental, and insurmountable, problem is that the processes of colonization and competition don't actually produce the patterns they were long and widely assumed to produce. It's just not true that sufficiently strong local species interactions will set an upper limit to local species richness that would not otherwise exist. Models with competition can indeed exhibit an upper limit to local species richness (Case 1990)—but so can models lacking any competition (J. Fox and Srivastava 2006). Indeed, an extremely simple mathematical model lacking any local species interactions whatsoever can be made to produce either linear or saturating local-regional richness relationships just by tweaking model parameters (J. Fox and Srivastava 2006; see also chapter 5 for a bit more discussion of this model). Conversely, linear local-regional richness relationships have been found in natural systems in which experiments show that competition sets an upper limit to local species richness (Shurin 2000; Shurin, Havel, et al. 2000).

Local-regional richness relationships are analogous to a medical diagnosis based on an incorrect understanding of which diseases cause which symptoms. Strong competition within local communities often doesn't produce a saturating local-regional richness relationship. And linear local-regional richness relationships have many possible causes, besides an absence of strong competition.

METHOD FOR INFERRING PROCESS FROM PATTERN 2: INFERENCE TO THE BEST EXPLANATION

Now let's turn to a second, quite different way to infer process from pattern, exemplified by plate tectonics.

As noted by James Brown (1995), Alfred Wegener and other 20th-century geologists correctly inferred that solid sections of Earth's crust, known as plates, move around—a phenomenon known as continental drift. Geologists inferred this without being able to directly observe the movement; it was too slow to measure with the technology available at the time. Nor could geologists experimentally manipulate the movements of the plates. Nor could they do more than speculate about the physical mechanism by which the plates might move.

Rather, geologists inferred continental drift because it's the only way to explain every one of several observations (Oreskes 1999; Frankel 2012). First, the coastlines of widely separated continents fit together like puzzle pieces. Their continental shelves fit even better, because coastlines experience more erosion than shelves do. South America and Africa provide the clearest example. Second, ancient geological features such as Caledonian mountain belts are found on separate continents, in the places you'd expect if those continents had been joined at the time those features formed. Third, fossils of the same extinct terrestrial and freshwater species are found in rocks of the same age on different continents. Fourth, iron-rich lava is magnetized in the direction of Earth's magnetic field at the time it hardens. Lavas of different ages are magnetized in different directions, implying that the land masses on which they're located have moved (the alternative possibility that the magnetic poles have moved can be ruled out on other grounds). Fifth, glacial deposits and glacial striations from the Permo-Carboniferous glaciation 300 mya are found on the southern portions of all Southern Hemisphere continents, implying either that those continents were joined near the South Pole at the time, or that the Permo-Carboniferous glaciation covered almost the entire Southern Hemisphere. The latter possibility is inconsistent with other lines of geological evidence.

Notice that the logic of the inference here is quite different from that of medical diagnosis. It's not that geologists considered various different alternative hypotheses, only one of which was consistent with all five "symptoms" identified by Wegener (Oreskes 1999). After all, until you develop the hypothesis of continental drift, it's not even clear that all five "symptoms" identified by Wegener are in fact "symptoms" of the same underlying cause.

Rather, the inference of continental drift is an example of what philosophers of science call "inference to the best explanation" (also called "abduction"; Douven 2021). Imagine you're faced with several facts that you can't explain. They might not even be related to one another at all, as far as you can tell. The only thing they seem to have in common is that they're all quite puzzling. In light of what you know about the world, it's quite surprising to you that any of these facts should be the case. Now imagine that someone suggests a single hypothesis that, if it were true, would explain all those facts—would make all of them unsurprising. The proverbial light bulb would turn on in your head. "Aha!" you would think. "So that explains it! It all makes sense now!" You infer the hypothesis is true not because alternative hypotheses don't fit the facts, but because there aren't any alternatives. The alternative is to leave all those puzzling facts unexplained, or perhaps to come up with a separate hypothesized explanation for each of them. For instance, you might explain fossils of the same terrestrial species on different continents by hypothesizing that those species crossed oceans on now-submerged land bridges. But that wouldn't explain the distribution of Caledonian mountain ranges, the shapes of continental shelves, and so on.

Inference to the best explanation is controversial. On the one hand, it's a mode of reasoning that people use all the time in their everyday lives (Douven 2021). Ecologists (including me) use it all the time too; we collect data, then propose explanations that seem to fit those data. But on the other hand, philosophers disagree on why inference to the best explanation works, when it does work (Douven 2021). And it doesn't always work. Inference to the best explanation can fail badly. Once you know about the patterns in your data, it's often pretty easy to invent some explanation that fits the patterns. At which point, it often feels very natural to turn around and claim the good fit as evidence for the hypothesized explanation. This is known as HARKing: *h*ypothesizing *a*fter the *r*esults are *k*nown. It's as if you shot a shotgun at the side of a barn, then painted a target around the largest cluster of pellets and declared yourself a sharpshooter ("Texas Sharpshooter Fallacy" 2024). You came up with the explanation after you saw that data, with the goal of coming up with an explanation that would fit the data. So of *course* your explanation is going to fit the data, whether it's correct or not! Just as you can't possibly miss a target that you paint after firing your shot. To be clear, it's totally fine to think up a hypothesis that might explain the patterns in your data—so long as you don't turn around and use those same patterns as evidence for your hypothesis.

HARKing is common in ecology. In an anonymous survey of a large random sample of ecologists, half of those surveyed admitted to HARKing (H. Fraser, Parker, et al. 2018). Had I taken that survey, I'd have admitted to HARKing too. I suspect that HARKing in ecology leads to many incorrect inferences to the best explanation.

The point here is not that researchers who engage in HARKing are bad scientists—they're not! The point is that science is hard. Inference to the best explanation is a popular and potentially powerful way to infer process from pattern. But it's prone to failure. It's easy to fool yourself, and others, into thinking you've figured out the explanation for some pattern in your data, when in fact you haven't.

METHOD FOR INFERRING PROCESS FROM PATTERN 3: CALCULATION BASED ON DETAILED BACKGROUND KNOWLEDGE

In light of Robert MacArthur's example of astronomy as a successful science based on inferring process from pattern, it's appropriate to discuss a third way of inferring process from pattern, exemplified by an example from astronomy.

The chemical composition of stars can be inferred from spectroscopy. Passing starlight through a prism to split it into its component wavelengths reveals a spectrum of both emission lines and absorption lines. When heated, some chemical elements and compounds emit light, at wavelengths that are unique to that compound. Those are the emission lines. Absorption lines appear when a hot gas passes through a cooler one, as when the hot gas from a star's interior passes through the cooler exterior gas. This cancels out emission lines the star would otherwise exhibit. We know all this from studies of the spectra of chemical elements and compounds here on Earth. When astronomers first began performing spectroscopy on starlight, they found absorption and emission lines characteristic of many elements present on Earth, but the elements present and their relative amounts seemed to vary tremendously among stars (fig. 4.8).

Figure 4.8. Stellar absorption spectra

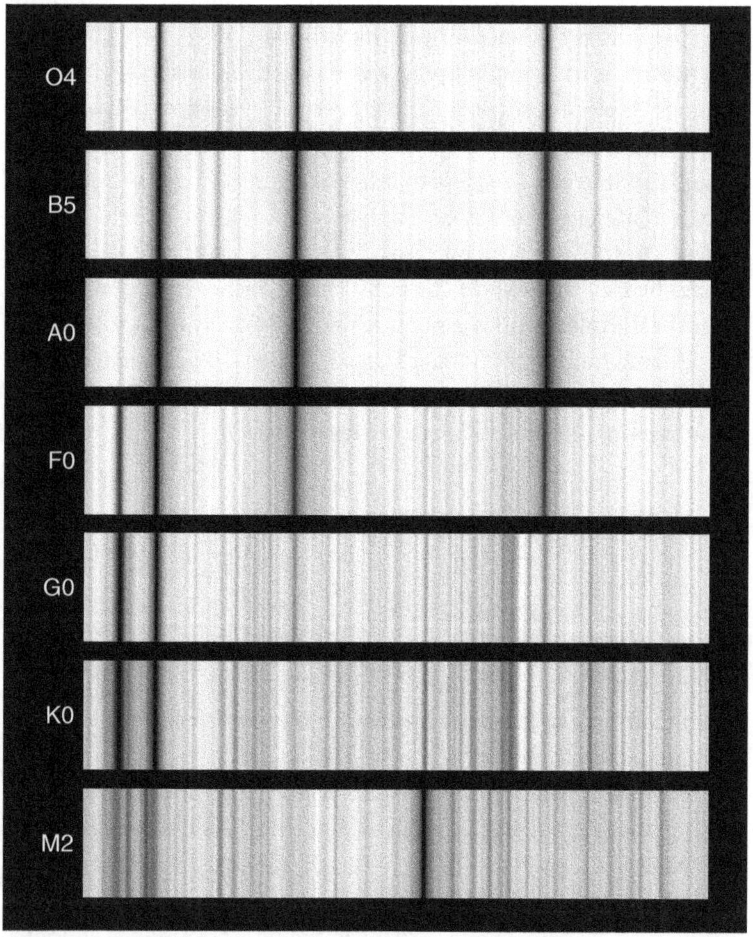

Stellar spectra are divided into classes corresponding to different temperatures: O, B, A, F, G, K, M (O-class stars are hottest, M-class coolest). The dark vertical bands in each spectrum are absorption lines. The lines are at different locations for each class, which might be thought to indicate differences in chemical composition. However, temperature and ionization alter absorption spectra, making them a misleading guide to stellar chemistry unless the effects of temperature and ionization are first corrected for.

Source: National Optical Astronomy Observatory.

However, astronomer Cecilia Payne-Gaposchkin (1925) realized that the raw spectrogram of a star is a misleading guide to its chemical composition. Different stars have different temperatures, which leads to differing levels of ionization of the stars' elements. That ionization in turn alters the stars' spectra; ionized atoms do not produce discrete absorption lines. Correcting for temperature and ionization using well-validated, quantitative physical theory (the Boltzmann equation and the Saha equation) reveals that all stars are approximately 90% hydrogen and 10% helium, with only trace amounts of other elements.

The logic of Payne-Gaposchkin's inference of process from pattern differs from both medical diagnosis and inference to the best explanation. Payne-Gaposchkin's inference relies on extensive, detailed background knowledge. We already know how various processes (here, stellar temperature, ionization, and stellar chemical composition) will combine to produce the pattern we observe (here, stellar absorption spectra). We can use that knowledge to subtract out, or adjust for, the effects of the processes that aren't of scientific interest (here, stellar temperature and ionization). This subtraction reveals the effect of the process of scientific interest (here, stellar chemical composition). That is, just because astronomers can't experiment on their study objects doesn't mean that experiments are thereby irrelevant to their work. Indeed, exactly the opposite: much of astronomy would be impossible without experimental data. Just because you can't experimentally manipulate whatever large-scale pattern you're studying doesn't mean that small-scale experiments are irrelevant to you.

Ecologists often have attempted to do the same thing Payne-Gaposchkin did: adjust observed data so as to subtract out the effects of some processes that aren't of scientific interest, thereby revealing the effects of processes that are of scientific interest. Usually, we do this by comparing the observed data to a "null model" describing the processes whose effects we want to subtract out. Chapter 5 discusses a successful example of this use of null models in ecology, to subtract out "sampling effects" from BEF experiments (Loreau and Hector 2001; J. Fox 2005).

But ecologists' attempts to use null models to subtract out processes that aren't of scientific interest often fail badly. There's a reason why the case I'm about to describe is one of the most oft-repeated cautionary tales in ecology.

Jared Diamond (1975) observed that some pairs of closely related bird species in the Papua New Guinea archipelago exhibited checkerboard distributions: any given island had one species or the other but never both. Diamond interpreted checkerboard distributions as evidence for interspecific competition, inferring that similar bird species couldn't coexist. But as critics pointed out, other processes besides interspecific competition could generate checkerboard distributions (Connor and Simberloff 1979). Perhaps the two species are best adapted to different environments, and each is found in the environment to which it is best adapted. Or perhaps their distributions relative to one another are just a matter of random chance.

In response, ecologists began to use randomized null models to infer whether the distributions of species among sites (e.g., occurrences of birds on islands) reflect interspecific competition (Gotelli and Graves 1996). The basic idea is to randomly shuffle the observed "species × sites matrix" (the data matrix recording which species occur at which sites). Random shuffling generates hypothetical data, in which the probability that a given species will occupy a given site is independent of whether any other species occupies the same site. Typically, the randomization is done under constraints to ensure that the randomized data retain other features of the observed data. For instance, you can randomize species' occurrences among sites under the constraints that each site ends up with the same number of species as it had in the observed data, and that each species occur at the same number of sites as it did in the observed data. From the observed data, one calculates a test statistic that summarizes the extent to which the species exhibit checkerboard-like distributions. One then compares the observed test statistic to the distribution of values from the randomized datasets. If species tend to co-occur less often than expected under the randomized null model, one infers that interspecific competition is at work.

This approach doesn't work, for two reasons. The one I'll focus on is the "Narcissus effect" (Colwell and Winkler 1984): the randomized data still retain some effects of interspecific competition. In particular, competitive exclusion means that interspecific competition prevents species from occupying sites they would've occupied in the absence of interspecific competition. Constraining the randomized data to retain each species' observed frequency of occurrence, and each site's observed species richness, therefore smuggles in effects of competitive exclusion. The problem here is analogous to the errors Payne-Gaposchkin would've made if she hadn't been able to use the Boltzmann and Saha equations to correct for effects of stellar temperature and ionization. You can't subtract out from your data the effects of processes other than the one of scientific interest, if you don't have sufficient background knowledge about what effects those processes would have on your data.

The other reason this approach doesn't work is analogous to a medical diagnosis that fails due to misunderstanding of which diseases cause which symptoms. Mathematical models and experiments show that interspecific competition doesn't invariably lead to any particular pattern in the distribution of species among sites, whether checkerboard or some other pattern (Hastings 1987; Ulrich, Jabot, and Gotelli 2017; Dallas, Melbourne, and Hastings 2019). Nor does lack of interspecific competition invariably lead to any distinctive patterns that couldn't possibly be produced by interspecific competition.

In the wake of the failure of randomized species × sites matrices as a method for inferring interspecific competition, ecologists turned to manipulative field experiments to test directly for interspecific competition (reviewed in Schoener 1983; Gurevitch et al. 1992). Notably, constrained randomization of species × sites matrices fails to recover strong interspecific competition revealed by experimental manipulations (Barner et al. 2018; Freilich et al. 2018).

Inference of interspecific competition isn't the only context in which ecologists have tried, and failed, to use constrained randomization of observed data to infer process from pattern. Consider research on the mid-domain effect.

If the locations of species' geographic ranges within a bounded geographic domain are randomly shuffled, the ranges will tend to overlap most toward the center of the domain. This has been termed a "mid-domain effect" (Colwell and Hurtt 1994). It has been argued that the mid-domain effect shows that peaks in species richness at tropical latitudes and at mid-elevation on mountainsides would have arisen in the absence of environmental gradients, simply due to "hard boundaries" such as coastlines and mountaintops (Colwell and Hurtt 1994; Colwell and Lees 2000). It has been further argued that the mid-domain effect provides a "null" model, in the sense that only deviations of observed species richness values from those predicted by the mid-domain effect require any explanation in terms of the causal effects of environmental gradients (Colwell, Rahbek, and Gotelli 2005).

These claims are controversial for several reasons. For instance, mid-domain effect models fit observed species richness patterns relatively poorly, especially patterns in two-dimensional space as opposed to one-dimensional transects (Currie and Kerr 2008). But for our purposes, the more important problem is that the mid-domain effect is a Narcissus effect. Randomizing the locations of species' ranges within a bounded domain, while maintaining the observed sizes of those ranges, fails to remove effects of environmental gradients on the sizes of species' ranges (Hawkins, Diniz-Filho, and Weis 2005; Zapata, Gaston, and Chown 2003, 2005). The importance of this is illustrated by simulations of models in which individual organisms give birth, die, and move around within a bounded domain with uniform environmental conditions, or with environmental variation that all species tolerate. Such models exhibit little or no mid-domain effect, because all species end up distributed throughout almost the entire domain (Rangel and Diniz-Filho 2005; Connolly 2005).

As with constrained randomization of species × sites matrices, the mid-domain effect is a failed attempt to correct for effects of processes other than the one of interest, without sufficient background knowledge about those other processes.

Many other examples could be given of unsuccessful attempts to infer process from pattern in ecology, all of which initially appeared promising and were taken up widely. These weren't bad ideas! Yet, all of them turned out to be analogous to making a medical diagnosis based on incomplete or inaccurate knowledge of which diseases cause which symptoms, or to calculating the chemical composition of a star without the Boltzmann and Saha equations.

- Inferring whether the species-abundance distribution arises solely from drift and migration, or whether other processes also play a role, by fitting alternative models to observed species-abundance distributions (Bell 2001; Hubbell 2001; McGill 2003; Wilson and Lundberg 2004; McGill, Etienne, et al. 2007; Hammal et al. 2015; see also chapter 7).

- Inferring whether competitive exclusion or "habitat filtering" governs the species composition of local communities, by observing whether co-occurring species are phylogenetically clustered or overdispersed (Mayfield and Levine 2010; Kraft et al. 2015).
- Inferring the dominant processes structuring metacommunities (networks of local communities linked by dispersal) by asking how much of the among-community variation in species composition is statistically associated with observed environmental variables versus spatial distance (T. Smith and Lundholm 2010; Gilbert and Bennett 2010; B. Brown et al. 2017; Barbier, Bunin, and Leibold 2025; for discussion, see chapter 5).
- Inferring whether foraging animals perform what are known as "Lévy walks" by observing whether the frequency distribution of their movement lengths follows a power law (Pyke 2015).
- Inferring the locations of good habitats for a species (i.e., locations where the low-density per capita growth rate will be high) from observations of where the species is present or absent (Tyre, Possingham, and Lindenmayer 2001).
- Inferring the contributions of ecological versus evolutionary processes to present-day biogeographic distributions from observed taxon age-range correlations (D. Warren et al. 2014).

WHY DO ECOLOGISTS KEEP SELECTING FOR INEFFECTIVE APPROACHES FOR INFERRING PROCESS FROM PATTERN?

Why do unsuccessful inferences of process from pattern run to type? Why do ecologists keep selecting for inferential approaches that fail for the same few reasons?

I can only speculate. Here's one notion: we're selecting for inferential approaches that can be applied easily to a wide range of cases. That's because we don't want just those rare case studies in which we ecologists can emulate medical diagnosis, or in which we correct for the effects of processes besides the one of interest. Ecologists generally don't want to infer process from pattern *just* in one particular case. We want to know if species' co-occurrence patterns *usually* reflect interspecific competition. If local species interactions *typically* are strong enough to limit local community membership. The cause(s) of *most* mid-domain peaks in diversity. And so on. It's challenging for ecologists to answer questions about what is typically the case, because doing so requires data from many different cases. To answer questions about what's typically the case, we need approaches that any ecologist can easily apply, in any study system of interest, without much customization. Which perhaps is why Hutchinson's ratios, constrained randomization of species × site matrices, local-regional richness relationships, randomizing species' observed geographic range locations, and other similar approaches all became temporarily popular. They could be applied to many different cases, could use already available or easily collected observational data, and didn't require tailoring to (or even much background knowledge about) the details of any particular case. But here's the problem: the same features that allow inferences from pattern to process to be attempted everywhere prevent them from working anywhere. It's not a coincidence that selecting for wide applicability also selects for ineffective applicability.

My own view is that ecologists should stop caring *quite* so much about what's typically the case. Don't stop caring about it at all; just care a bit less. Which doesn't mean giving up on all possibility of generality, by the way. In chapter 7, I argue that there are several different senses of "generality" in ecology, of which "What's typically the case?" is only one. Notably, generality in other senses can be, and has been, pursued using a diverse range of approaches and combinations of approaches.

Failure of Selection Due to Lack of Variation

As noted earlier in this chapter, there's one more reason why selection effects can fail to generate a positive effect of biodiversity on ecosystem function: lack of initial diversity. Biodiversity can't improve ecosystem function via a selection effect if there's no biodiversity to begin with. Analogously, selection effects in science can't operate if there aren't diverse ideas or approaches to select among. Lack of diversity in ideas and approaches is an important enough, and distinct enough, problem that it deserves its own chapter (chapter 6). The problem of how to harness the diversity of ideas, goals, and approaches is a different problem than the problem of how to get diverse ideas, goals, and approaches in the first place.

Conclusions

I don't think it's a coincidence that the case in which selection works reasonably well—peer review—is a case in which it's designed to work reasonably well. The scientific peer-review system has numerous features designed to reduce biases and randomness in the selection process. These days in ecology, it's common (though not universal) for reviewers to be blinded to who wrote the paper. Each manuscript that goes out for review receives multiple independent reviews rather than just one. Those reviews come from experts with no conflict of interest with the authors. Reviewers receive instructions as to what features of the manuscript to select for. And so on. Peer review obviously isn't perfect, and there's certainly room for further improvement. But overall, it does a difficult job pretty well.

In contrast, cases where selection doesn't work as well are cases in which it's not designed to work as well. Failure to test alternative scientific hypotheses, publication bias, presentation bias, and repeated selection of ineffective methods for inferring process from pattern—all are cases where those doing the selecting have cogent reasons to select for the wrong things, or not to do any selection at all. Without much in the way of countervailing reasons.

I don't know how much can be done about this. Hopefully, awareness of cases in which selection works well, and cases in which it works less well, will help ecologists get better at selection in the future. There certainly are plenty of cases in the history of ecology in which we've improved at selection. Nobody any longer tries to, say, infer interspecific competition from Hutchinson's ratio, or from constrained randomization of species × site matrices. Those ineffective approaches were eventually selected against. The hope is that, in the future, selection against ineffective approaches can proceed a bit more quickly.

5 *Diverse Tools for Diverse Jobs: The Many Uses of Mathematical Models*

In the previous two chapters, we talked about ways in which diverse scientific ideas and approaches can accelerate scientific progress, via effects analogous to selection and complementarity effects in BEF experiments. In the course of those discussions, I've referred repeatedly to experiments like George Sinclair's original BEF experiment, which asked how increasing the species diversity of a plant community affects primary productivity.

But primary productivity is just one ecosystem function among many. As noted in chapter 2, ecosystems can be said to perform many different functions besides primary production. And many of those functions are performed by species that aren't plants. BEF researchers sometimes refer to these facts as "multifunctionality" (Hector and Bagchi 2007). But rather than use that bit of jargon, I'll use the more familiar phrase "diverse tools for diverse jobs." A third way biodiversity can promote ecosystem function is by having different species that perform different functions. Analogously, in scientific research we often need to use different tools to do different jobs.

That broad point is so obvious, I'm sure you don't need me to convince you of it. But merely recognizing that we need diverse tools for diverse jobs doesn't necessarily allow us to recognize the full diversity of tools available to us, or the full diversity of jobs they're capable of doing. In this chapter, I'll talk about a comparatively little-used tool in ecology: mathematical modeling. I'll suggest that mathematical modeling is little used at least in part because it's not widely recognized as being a diverse set of tools, capable of doing a diverse range of different scientific jobs.

A Very Brief History of Mathematical Models in Ecology

Philosopher George Santayana (1905, 284) famously wrote that "those who cannot remember the past are condemned to repeat it." But as the history of mathematical modeling in ecology illustrates, sometimes you're condemned to repeat the past even if you do remember it.

Way back in 1935, Charles Elton, author of the first textbook of animal ecology, reviewed one of the first attempts to apply mathematics to ecology, A. J. Lotka's *Théorie analytique des associations biologiques: Part 1: Principes*. Elton (in)famously found the book a difficult read, writing of Lotka:

> Like most mathematicians, he takes the hopeful biologist to the edge of a pond, points out that a good swim will help his work, and then pushes him in and leaves him to drown. (Elton 1935, 149).

But Elton also found the book well worth the effort of getting to grips with:

> Behind these ordinary ecological investigations, concerned with organisation of observers, census techniques, and all the paraphernalia of field surveys and the laboratory and museum activities that go with them, lies this basic work upon populations treated by Lotka. He seeks to create a system of mathematical formulae which will guide the ecologist in his population studies. . . . The importance of the method is this: if we know certain variables, mostly desired by ecologists and in some cases already determined by them, we can predict certain results which would not normally be predictable or even expected by ecologists. The stage of verification of these mathematical predictions has hardly begun; but their importance cannot be under-estimated, and we look forward to seeing the further volumes of Lotka's studies. (149)

At least, Elton found the book well worth getting to grips with in 1935. His attitude toward mathematical modeling seems to have cooled later in life. Mathematical biologist Jim Murray met Elton in the 1970s. Their brief encounter went amusingly poorly (J. Murray 2001, 36):

> When I expressed an interest in animal ecology and mentioned his seminal work on the data from the Hudson Bay Company he started to talk enthusiastically about oscillatory behavior in populations. When I said I was a mathematician working in biology there was a notable cooling of his enthusiasm and he said, "Oh, you're one of them" and added, "I thought you were somebody else."

Those anecdotes about Charles Elton are representative of early 20th-century ecologists' reactions to mathematical modeling: a mixture of curiosity, skepticism, and incomprehension (Kingsland 1995). The more things change, the more they stay the same, as the old saying goes. Every few years, an ecologist adds to the growing list of papers and blog posts lamenting the lack of integration between theoretical and empirical ecology, diagnosing the reasons for the lack of integration, and suggesting remedies (Caswell 1988; Kareiva 2014; Grimm 1994; Ellner 2006; Fawcett and Higginson 2012; Scheiner 2013; J. Fox 2013a; Haller 2014; Rossberg et al. 2019; Servedio 2020; Grainger et al. 2022; Ou et al. 2022). This isn't just a matter of people complaining for no reason either; there's plenty of data showing how disconnected mathematical modeling is from other research approaches in ecology (Caswell 1988; Scheiner 2013; Haller 2014; Servedio 2020; see also chapter 1).

No doubt there are many reasons for this disconnect. "Love of math" is a rare motivation for studying ecology. Undergraduate ecology courses typically don't involve much math beyond basic statistics. The typical mathematical ecology paper is aimed at readers who already know math, often making it a difficult read for others. And so on.

But perhaps another reason is lack of appreciation for the diversity of ways mathematical models can be used in ecology. There are many, many ways to use mathematical models to learn about the world—many of which are rarely used in ecology. But it's hard to learn about those myriad uses if you don't know much math to begin with. It all just blurs together as "math." So in the hopes of providing some inspiration,

I'll review some of the many uses of mathematical models in ecology. "Math" isn't just one tool for one job ("modeling"). Rather, math is an entire toolbox of diverse tools for diverse jobs.

1. USE A PHENOMENOLOGICAL MATHEMATICAL MODEL TO DESCRIBE RELATIONSHIPS AMONG VARIABLES

By a "phenomenological" mathematical model, I mean a mathematical model that describes relationships between variables without making any claims about why those relationships hold. The most common phenomenological models in ecology are statistical curve-fitting models, relating the mean of one or more dependent variables to the values of one or more predictor variables. Here are some examples (some of which are well established, others more controversial):

- The power law relating the number of species in an area, S, to the size of the area, A: $S = cA^z$ where c and z are constants (reviewed in Rosenzweig 1995).
- Allometric scaling models relating various ecological variables (e.g., metabolic rate, population density, size of home range) to adult body size: $Y = aM^b$ where Y is the ecological variable of interest, M is the mean adult body mass of the species, and a and b are constants (Kleiber 1932; R. Peters 1983).
- Taylor's (1961) power law, which says that the variance of a species' abundance over space or time is proportional to its mean abundance raised to some power, usually between 1 and 2 (reviewed in Xiao, Locey, and White 2015).
- The energetic equivalence rule, stating that the energy consumption of a population is independent of the mean body size of the organisms making up the population (Damuth 1991).

Mathematical descriptions of the relationships between variables are useful for various reasons. Here are a couple:

- You can use them in meta-analyses and other comparative analyses. You can't do a quantitative comparative analysis without first having some numbers to compare. For instance, Rosenzweig

(1995) documents variation in the exponent *z* of the species-area curve among different kinds of datasets (e.g., oceanic islands of different areas vs. nested sampling areas on continents). Rosenzweig (1995) discusses methodological and ecological explanations for those differences. Another kind of comparative analysis involves identifying outliers—observations that deviate from the usual relationship between variables and so require an explanation.
- You can interpolate and extrapolate them to make predictions. For instance, several researchers have extrapolated species-area curves to predict how many species will be driven extinct by a given amount of habitat loss (He and Hubbell 2011, which explains why those predictions overestimate extinction risk).

Finally (and to my mind, most powerfully), you can use a phenomenological model as a starting point for a mechanistic model. The phenomenological mathematical description of how two or more variables are related gives you something to explain. You've summarized the relationship between two variables using one or more numbers. Now you can ask: *why* do those numbers take on the values they do, rather than some other values?

For instance, the allometric exponent *b* relating metabolic rate to body size typically has a value close to 3/4 (G. West, Brown, and Enquist 1997; Banavar et al. 2010). A large body of theoretical and empirical work proposes and tests hypotheses explaining why *b* takes on a value close to 3/4, as opposed to some other value such as 2/3 (e.g., G. West, Brown, and Enquist 1997; Banavar et al. 2010).

As another example, consider ectotherm development rate. For ectotherms, development rate (the inverse of the time required to complete development) is linearly related to temperature, within the range of temperatures that developing organisms normally experience. Development rate equals zero below some critical threshold temperature (Pauly and Pullin 1998, Charnov and Gillooly 2003). That is, development requires some species-specific, approximately constant number of degree days above the critical minimum temperature (Charnov and Gillooly 2003). The required number of degree days in turn depends on adult body size (Charnov and Gillooly 2003). The relationships among those variables can be described with phenomenological linear regression models. The form of those phenomenological regression models in turn suggests a more mechanistic explanatory model (Charnov and Gillooly 2003). That more mechanistic model predicts an exponential relationship between development rate and temperature—but an exponential relationship that is approximately linear over the range of temperatures typically experienced by any given species (fig. 5.1).

As an aside, it is worth noting that all mathematical models in ecology are "phenomenological," in the sense that every parameter, variable, and function can be viewed as an approximation to or summary of some more-complicated underlying reality that the model treats as a black box. For instance, simple predator-prey models such as the Rosenzweig and MacArthur (1963) model have a parameter called the "conversion efficiency." Conversion efficiency is the number of predators born per prey consumed, or units of predator biomass produced from consumption of

Figure 5.1. Expected ectotherm development rate (corrected for body size and food quality) as a function of temperature

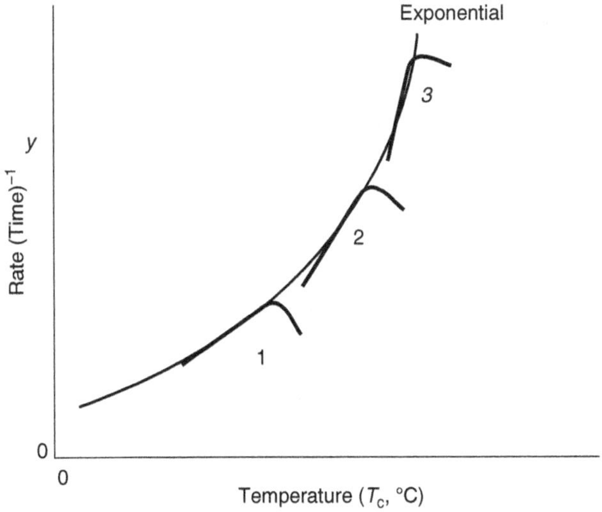

A given species will occupy only a limited range along the expected exponential curve; three hypothetical species are shown. Each species exhibits an approximately linear relationship between development rate and temperature.

Source: Modified from Charnov and Gillooly (2003).

one unit of prey biomass. The conversion efficiency parameter summarizes the consequences of all the underlying physiological, reproductive, and developmental processes that go into converting consumed prey into new predators. As another example, describing a process as "stochastic" or "random," and modeling it using some probability distribution, is a phenomenological way of summarizing the net effect of many complicated underlying deterministic processes, about which we know little or nothing. As a third example, Glazier (2014) argues that purportedly mechanistic models proposed to explain allometric scaling are themselves better regarded as phenomenological summaries of complicated underlying biological mechanisms. Much of the art of modeling involves exercising good judgment about when and how to summarize the consequences of some underlying mechanism, rather than modeling that underlying mechanism explicitly.

2. FORECASTING

Another use of mathematical models is for forecasting. I'll walk you through an example I lived through. Back in 2001, I was a postdoc in Britain during an outbreak of foot-and-mouth disease.

Foot-and-mouth disease (FMD; also called hoof-and-mouth disease) is a viral disease infecting cattle, sheep, pigs, and other cloven-hoofed animals (Haydon, Kao, and Kitching 2004). It is highly contagious, and spreads by both direct contact and airborne transmission (even over distances of more than a kilometer). The symptoms include blisters on the feet and mouth, hence the name "foot-and-mouth disease." Infection is rarely fatal; most infected animals eventually recover. But infected animals become lethargic, stop eating, stop gaining weight, and stop producing milk. Infected animals also can't be sold for export. British livestock farms had been FMD-free since 1967, until an FMD outbreak started in February 2001. The outbreak may have originated from infected pig feed imported into Britain.

Prior to 2001, FMD outbreaks in Britain were controlled by slaughtering infected animals within 24 hours, and also slaughtering animals at high risk of having been infected via close contact with an infected animal on the same farm (Haydon, Kao, and Kitching 2004). This approach almost always succeeded in preventing FMD from spreading;

it failed only once, in 1967 (Haydon, Kao, and Kitching 2004). It failed again in 2001, for a combination of reasons (Haydon, Kao, and Kitching 2004). The rapid spread of FMD in the face of the apparent failure of the usual control measures led the British government to consider alternative control measures. This immediately raised a challenging question: what control measures, exactly?

To help answer that question, the government turned to mathematical modeling. Three independent research groups were tasked with building mathematical models of FMD spread in Britain (Ferguson, Donnelly, and Anderson 2001; Haydon, Kao, and Kitching 2004). The models were used to forecast the future course of the epidemic under a "business as usual" scenario, and to forecast the efficacy of various novel control strategies. The three models used different mathematical approaches and made somewhat different assumptions, while also sharing some assumptions in common. All agreed that "business as usual" would result in a massive epidemic that would decimate the entire British livestock industry (fig. 5.2). Instead, the government decided to metaphorically burn the village in order to save it. The mathematical models predicted that the outbreak could be brought under control most quickly, as well as minimizing the total number of livestock that would eventually need to be slaughtered, via a new control measure: preemptive slaughter (fig. 5.2). The government called in the military to assist with rapidly culling all livestock on farms within 3 km of, or contiguous with, an infected farm, whether the animals were infected or not. The growth rate of the epidemic soon slowed, although a long "tail" of new cases continued for many months before the outbreak was over. In the end, millions of animals—a substantial minority of all livestock in Britain—were slaughtered, the majority of which were slaughtered preemptively.

Preemptive slaughter was and remains very controversial. It was chosen in part for logistical and political reasons. In an urgent and uncertain situation, there is logistical and political value in clear public policies that are relatively straightforward to implement and to explain to the public. But preemptive slaughter was also chosen because mathematical modeling was the only way to forecast the effects of different management options, on the basis of the limited and imperfect information available.

Figure 5.2. Forecasted FMD case incidence in the model of Ferguson, Donnelly, and Anderson (2001) under different management scenarios

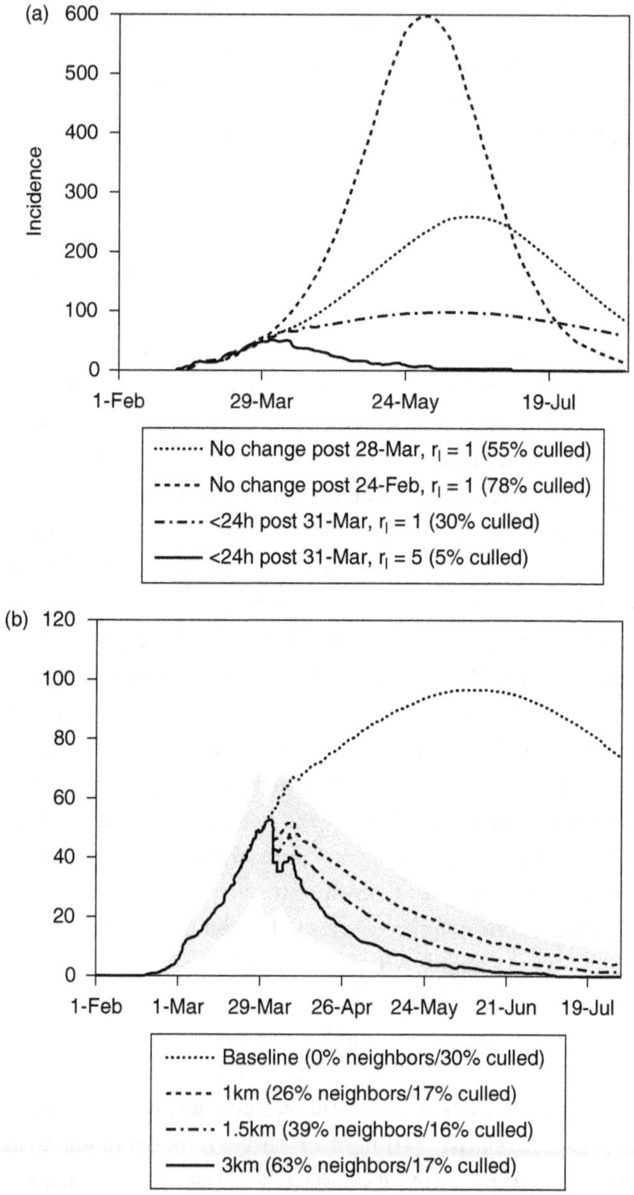

(a) Business-as-usual forecasts, and forecasts imagining that animals on infected farms could be slaughtered within 12 hours of discovery of an infection. (b) Forecasts imagining "ring culling" of all animals on all farms within a given radius (1 km, 1.5 km, or 3 km) of an infected farm, beginning on April 1, 2001.

Source: Modified from Ferguson, Donnelly, and Anderson (2001).

3. CHECK YOUR LOGIC

Often in science, we know, or think we know, something about the world, and we want to work out the implications. If this is true, what else has to be true? Or maybe we don't know it, but we assume it as a working hypothesis. Assuming *x* is the case, would I expect to observe *y*?

One important use of math is to check your logic: help you figure out the implications of what you know, or think you know, or assume. Everyone needs that help. Many things that ecologists study are very difficult to reason about verbally, for many reasons:

- feedbacks (*a* affects *b* and vice versa)
- nonlinearities (the effect of *a* on *b* depends nonlinearly on the amount of *a*, as in cases of "too much of a good thing")
- stochasticity (the effect of *a* on *b* is at least partially a matter of chance, like the roll of a die)
- time lags (*a* affects *b*, but the effect takes some time to manifest)
- lengthy chains of cause-effect relationships (*a* affects *b*, which affects *c*, which affects *d* . . .)
- multiple causal drivers that interact in various ways (the effect of *a* on *b* depends in some way on *c*; the effects of *a* and *b* on *c* cancel one another out; other possibilities)

For all those reasons, it's difficult to work out the implications of what you know, or think you know, or assume. It's also difficult to recognize when you've made a mistake in working out those implications.

Further, it's not always clear what you even know, or think you know, or assume. Many important ecological concepts are defined only fuzzily (see chapter 8). In order to work out the implications of what you know, or think you know, or assume, you can't be too vague. What *exactly* is it that you know, or think you know, or assume?

Math can help. Building a mathematical model forces you to be precise about what you know, or think you know, or assume. Analyzing the model allows you to work out the implications. Let's look at some examples of this use of mathematical models.

3a. Make Your Assumptions Clear

Robert May used a very simple mathematical model to show that a species-rich community, in which each species interacts with many others (e.g., as a competitor, predator, prey, or mutualist), has vanishingly low odds of possessing a stable equilibrium to which species' densities will tend to return in the long run. May's (1972) model is far too simplified to describe the dynamics of any real ecological community with any precision. But it rightly undermined the then-common intuition, tracing back at least as far as Charles Elton (1958), that "diverse" and "complex" communities should be more "stable." In May's model, the opposite is the case: the higher the diversity (number of species) and complexity (fraction of species in the community with which the typical species interacts), the lower the odds of stability. Subsequent diversity-stability theory explored a wide range of mathematical models of different ecological scenarios, and different definitions of "diversity" and "stability" (Goodman 1975; de Ruiter, Neutel, and Moore 1995; McCann, Hastings, and Huxel 1998; McCann 2000; Yachi and Loreau 1999; Doak et al. 1998; Tilman, Lehman, and Bristow 1998; Ives, Klug, and Gross 2000; Cottingham, Brown, and Lennon 2001; Allesina and Tang 2012; Grilli et al. 2017; Y. Wang et al. 2019; Lamy et al. 2021). Whether or not diversity or complexity promotes stability turns out to be sensitive to *exactly* how you define "diversity," "complexity," and "stability," in ways that would have been impossible to even recognize, much less study, without the aid of mathematical models.

3b. Test Your Pre-mathematical Intuitions

One of the most important—and in my view, underrated—uses of mathematical models in ecology is to check whether our pre-mathematical intuitions are actually correct. Mathematical modeling doesn't (just) generate testable hypotheses. Sometimes, the mathematical model *is* the test—of the hypotheses generated by our pre-mathematical intuitions.

That last sentence might sound wrong. Surely we can test hypotheses only with data, right? No, actually. Before it's worth testing whether a hypothesis is false by checking its predictions against the data, we need to test if it could *possibly* be true or false.

To see this, let's remind ourselves why hypothesis testing works. We start with some set of claims about some actual or hypothetical world. Depending on what scholarly field you were trained in, you might call these claims your "model" (mathematical or otherwise), your "assumptions," your "premises," or some other term. From those claims, various implications follow logically, which you could call "predictions" or "deductions" or some other term. If the assumptions are true, the predictions *have* to be true, because they're logical implications of the assumptions. For instance, if Socrates is a man, and all men are mortal (assumptions), then it follows logically that Socrates *must* be mortal. Which means that if the predictions turn out to be false, then we can infer that at least one of the assumptions *must* be false.

But the logic of hypothesis testing falls apart if the purported predictions *don't* follow from the assumptions. The technical philosophical term for this is an "invalid" deduction. If the predictions of our model are logically disconnected from its assumptions, then we can't learn from the model. There's no point in checking whether the predictions are false or true (or approximately true, or true in some contexts, etc.), because that would give us no reason to reject the (logically unconnected) assumptions. If we assume/claim/propose that Socrates is a man, and that all men are mortal, and then deduce/conclude/predict that Socrates likes ice cream, our logic is faulty. Which means there's no point trying to find out whether Socrates actually likes ice cream. Maybe he does, maybe he doesn't, but the truth or falsehood of the conclusion has nothing to do with the truth or falsehood of the premises. Our argument is a non sequitur. Nor is there any point in checking whether the assumptions are true (or approximately true, or true in some contexts, etc.), because that wouldn't tell us anything about the truth of the (logically unconnected) predictions.

That's a deliberately silly example. The faulty logic is laughably obvious. But it's hard to reason about the complicated things ecologists study. So it's not always obvious when our nonmathematical models are non sequiturs.

As an example of a verbal model with conclusions that don't actually follow from its premises, consider G. Evelyn Hutchinson's (1961) famous proposal that intermediate frequencies of environmental change can prevent competitive exclusion. Competitive exclusion is prevented, Hutchinson claimed, because different species compete best in different environments. Every time environmental conditions change, the identity of the dominant competitor changes. This prevents competitive exclusion so long as the environment never remains unchanged long enough for one species to exclude the others. But neither can the environment change states too often, since in that case the competing species would "average over" the rapid environmental fluctuations, and whichever species was the best competitor on average would exclude the others.

This verbal argument is very intuitively appealing. I bet it sounds right to you! It certainly sounded right to me when I first learned it from my undergraduate ecology textbook (Begon, Harper, and Townsend 1990, 749–50). So it's no criticism of Hutchinson that this argument is a non sequitur. Hutchinson's premises actually imply that the winning species is always whichever one is the best competitor on average—that is, averaging over the fluctuating environmental conditions (Chesson and Huntly 1997; J. Fox 2013b). If you do the math, you discover that, on Hutchinson's premises, the frequency with which the environment fluctuates is irrelevant to the competitive outcome (fig. 5.3). Now, there are other models, with different assumptions, in which environmental fluctuations of intermediate frequency do enable species coexistence (Chesson and Huntly 1997; J. Fox 2013b; Klausmeier 2010). But that doesn't rescue Hutchinson's model. The fact that a valid model happens to lead to the same conclusions as an invalid model doesn't validate the invalid model.

There are many other examples of this use of mathematical models in ecology, but readers may also wish to look beyond ecology for additional examples. Otto and Rosales (2020) discusses this use of mathematical models in an evolutionary context.

3c. Discover a New Possibility That You'd Never Have Thought of Otherwise

Sometimes, in the course of using a mathematical model to check your pre-mathematical intuitions, you stumble across new possibilities that would never even have occurred to you if you hadn't done the math.

Figure 5.3. If the relative fitnesses of competing species depend solely on environmental conditions, the frequency with which conditions change is irrelevant to the competitive outcome

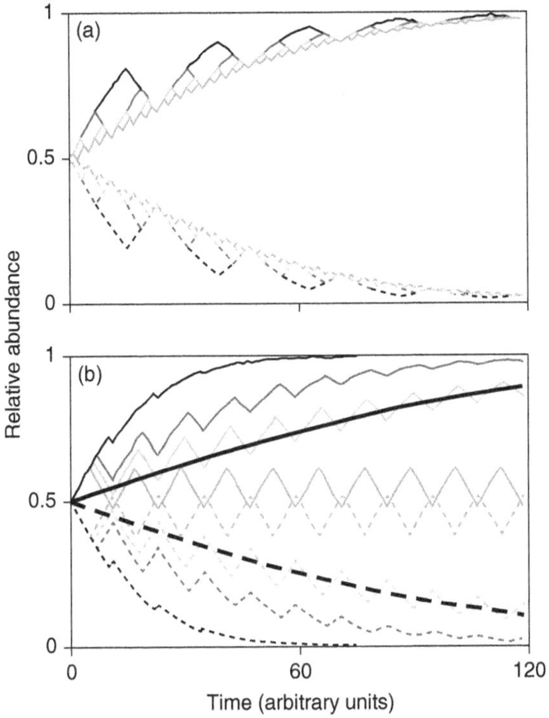

(a) Dynamics of relative abundance of two competing species (*broken and unbroken lines*) for different frequencies of environmental change (*different shades of gray*). The environment alternates between a state in which species 1 (*solid line*) has a relative fitness of 1 and species 2 (*broken line*) has relative fitness of 0.9, and a state in which relative fitnesses are reversed. Species 1 is favored 2/3 of the time in every case, and so the rate of exclusion is the same in every case. (b) As in panel a, but now different shades of gray indicate scenarios in which species 1 is favored different fractions of the time, ranging from 11/12 of the time (*thin black lines*) to 1/2 the time (*lightest gray lines*). The rate of exclusion slows as species 1 is favored less often, and when each species is favored 1/2 the time they coexist via neutrally stable oscillations around their initial relative abundances. The heaviest black lines show dynamics for a scenario in which species 1 is slightly favored at all times, so that it has the same fitness advantage on average as when it is favored 7/12 of the time (*lightest gray lines*).

Source: J. Fox (2013b).

Back in the early 2000s, my then-postdoc David Vasseur and I were interested in the factors that generate correlations in the fluctuations in the abundances of different species. In particular, it's widely thought that competing species will exhibit negative correlations—when one increases in abundance, its competitor will decrease, and vice versa (e.g., McCann, Hastings, and Huxel 1998; Loreau and de Mazancourt 2008). Such "compensatory" dynamics should stabilize the total abundance or biomass of the entire community of competitors, in the face of disturbances or environmental fluctuations: if one species goes down when another goes up, their total abundance or biomass won't change much. But presumably (David and I thought), such compensatory dynamics have their limits. Sufficiently severe disturbances or environmental fluctuations should overwhelm any tendency for competition to produce compensatory dynamics. For instance, during a major drought, you'd expect all plant species to decline in abundance, regardless of whether they're competing. But just how severe do the disturbances or environmental fluctuations have to be in order to overcome the tendency of competition to generate compensatory dynamics?

To answer that question, we started with a published mathematical model of competing species that had already been shown to exhibit compensatory dynamics (McCann, Hastings, and Huxel 1998). In that model, competing species exhibited compensatory fluctuations in abundance generated by their interactions with their shared resources and predators. David and I incorporated environmental fluctuations into the model. In good environments, all the competitors would experience a low mortality rate; in bad environments, a high mortality rate. The environment fluctuated randomly over time between better and worse environments. It turned out that even pretty small environmental fluctuations would make the competing species exhibit positively correlated fluctuations in abundance, rather than negatively correlated (i.e., compensatory) fluctuations.

We were all set to write a paper about how compensatory fluctuations generated by interspecific competition were easily overwhelmed by environmental fluctuations to which all species respond in the same way. But before we started writing, I suggested that we also look at the case of environmental fluctuations to which species respond in opposite ways—that is, a good environment for one species is a bad environment for its competitor, and vice versa. I thought this was a formality, just a matter of dotting our i's and crossing our t's. Obviously, environmental fluctuations to which competing species respond in opposite ways would reinforce, rather than overwhelm, the compensatory dynamics generated by competition.

Except it turned out it wasn't obvious. It wasn't even true! I vividly remember David coming to my office and telling me that the competing species were still exhibiting synchronous fluctuations in abundance, even though they were now affected in opposite ways by the environmental fluctuations. I told David that this made no sense. Surely he must've made a mistake in the computer code he wrote to simulate the model. David agreed that we needed to make sure there was no mistake, and he recoded the entire model from scratch. It still behaved the same way. There was no mistake. The problem wasn't with the model; the problem was with our intuitions.

We'd stumbled across something new: the same competition that generates compensatory dynamics in a constant environment, generates synchronous dynamics in response to random environmental fluctuations. Here's the gist (fig. 5.4; refer to Vasseur and Fox 2007 for the technical details I'm about to gloss over). Even if competing species respond in opposite ways to environmental fluctuations (i.e., a good environment for one is bad for the other), their opposing responses aren't likely to offset one another perfectly. So total competitor abundance will change a bit in response to an environmental change. That will then cause the abundances of shared resources and predators to change. For instance, if total competitor abundance goes up a bit, because one competitor increases a bit more than the other decreases, resource abundance will go down, and predator abundance will go up. This in turn will drive the abundances of *both* competitors down—a synchronous fluctuation. Repeated environmental fluctuations constantly perturb competitor growth rates, leading to perturbations to resource and predator abundances, to which both competitors respond in the same way. So in a fluctuating environment, competing species fluctuate synchronously in part *because of* (rather than despite) the fact that they compete via shared resources and shared predators.

Figure 5.4. Asynchronous fluctuations in competitor abundances, due to competition mediated by shared resources and shared predators

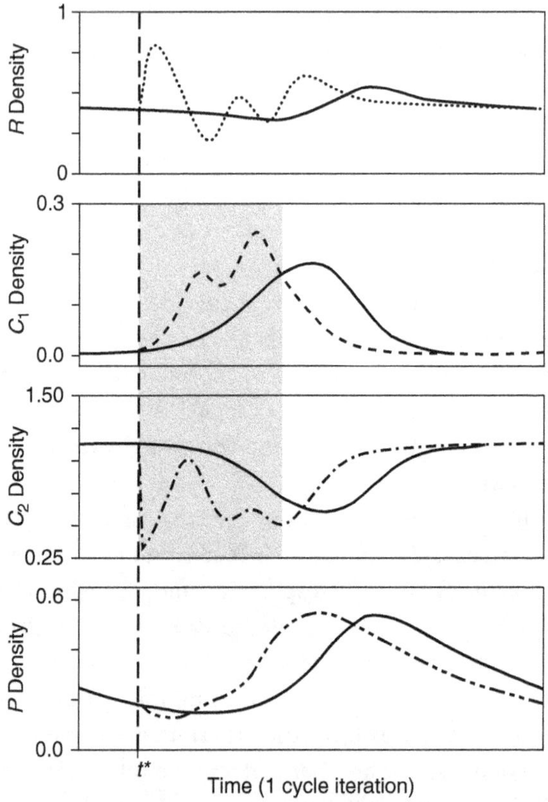

Competing consumer species C_1 (*dashed line*) and C_2 (*dash-dotted line*) fluctuate synchronously in response to a brief environmental change at time *t**. The environmental change slightly increases the density of C_1 and drastically decreases the density of C_2. In response to these changes in consumer densities, the resource *R* for which both consumers compete increases in density (*dotted line*), and the predator *P* that attacks both consumers decreases in density (*dash-double-dotted line*). Both consumers increase in density in response to the changes in *R* and *P*, and continue to fluctuate synchronously over the period of time highlighted by the gray rectangle. The solid curve in each panel gives the population dynamics that would've occurred in the absence of any environmental change. Note that C_1 and C_2 fluctuate asynchronously in the absence of environmental fluctuations—when one increases in density, the other decreases, and vice versa.

Source: Modified from Vasseur and Fox (2007).

3d. Use a Simple Mathematical Model to Define a Limiting Case

You aren't limited to using math to check your pre-mathematical logic. You can also use one mathematical model to check the logic of another mathematical model. One way to do that is by taking limits. Recall from calculus that "taking a limit" means studying the behavior of some equation or other mathematical object, as the value of one of its parameters is dialed up or down toward some value (usually, the highest or lowest possible value, often zero or infinity). Failure of any mathematical model to converge to appropriate limits raises the suspicion that the model badly misdescribes the things it models, even far away from those limits. Failure of any model to converge to appropriate limits also interferes with our ability to learn from the model. As discussed elsewhere in this chapter, many ways of learning from mathematical models involve comparing different models to one another—for instance, comparing a model that includes some factor to an otherwise equivalent model lacking that factor. But what if the behavior of the model that includes the factor of interest does not approach that of the model lacking the factor of interest, as the strength or magnitude of the factor approaches the lower limit of zero? Then we should conclude that the factor has been modeled badly. The model that includes the factor is false in a way that inhibits learning rather than aiding it.

As an example, consider ratio-dependent predator-prey models. A ratio-dependent predator-prey model is one in which predator per capita feeding rate is proportional to the ratio of prey density to predator density (reviewed in Abrams 2015). Ratio-dependent predator-prey models imply that the feeding rate of an individual predator approaches infinity in the limit as predator density approaches zero (Abrams 2015). That's impossible, obviously. Of course, predator-prey models with prey-dependent functional responses (i.e., predator per capita feeding rate depends only on prey density, not predator density) are themselves sometimes false, by virtue of incompleteness. Prey-dependent functional responses omit the fact that predator per capita feeding rates do sometimes depend on predator as well as prey densities—for instance, due to predators stealing one another's food, or fighting with one another over food (Abrams 2015). But prey-dependent functional response models nevertheless define the limit to which any functional response model including predator dependence should converge when predator densities are sufficiently low. That ratio-dependent functional response models fail to converge to this limit is a strike against them. Barraquand (2014) discusses other limits to which ratio-dependent predation models should converge but do not.

As a second example of this use of mathematical models, Alexander Strang, Karen Abbott, and Peter Thomas (2019) considered how to incorporate stochasticity—random variation in per capita birth and death rates—into simple models of population growth. Specifically, they considered a simple stochastic population growth model with an Allee effect: mean per capita population growth rate increases with population density below some threshold density, for some unspecified biological reason (for instance, maybe it gets easier to find a mate as population density increases). Above the threshold, mean per capita population growth rate decreases with increasing population density, because of intraspecific competition. Now, imagine reducing the threshold density below which the Allee effect operates. As you reduce the threshold toward the lower limit of zero, you'd expect the behavior of a model with an Allee effect to approach that of an equivalent model without an Allee effect. After all, if you dial the threshold density all the way down to zero, you eliminate the Allee effect entirely. But in fact, that's not what happens. Rather, as the threshold density is reduced toward zero, the expected persistence time of a population with an Allee effect increases more and more, relative to the expected persistence time of a population with no Allee effect. This makes no sense. As you make models with and without an Allee effect more *similar* (by dialing down the threshold density below which the Allee effect occurs), their dynamics become more *different*? Strang, Abbott, and Thomas traced this nonsensical limiting behavior to the way in which simple stochastic population growth models incorporate an Allee effect. The usual way of introducing an Allee effect below some threshold population density has the unintended side effects of stabilizing population dynamics above the threshold density, and stabilizing them more strongly as the threshold density is reduced toward the lower limit of zero. Oops! Strang, Abbott, and Thomas showed how to avoid nonsensical limiting behavior by first explicitly describing stochastic individual demography. That is, they first specified the probability that an individual will die, mate, and so on, during a specified time interval, and how those probabilities vary with population density. They then derived the stochastic population growth model implied by those explicit assumptions about individual-level stochasticity. The alternative approach of starting with a deterministic population growth model with an Allee effect, and then "bolting on" stochasticity, amounts to making unintended implicit assumptions about individual demography. Those implicit assumptions are very unrealistic, and they create nonsensical population dynamics.

3e. Combine Different Bits of Empirical Information and Work Out Their Implications

Often in ecology, we have various bits of empirical information that are relevant to some scientific hypothesis. Some bits might seem consistent with the hypothesis, others inconsistent. But it's hard to tell for sure, because it's hard to see the complete picture. It's hard to tell if all the bits of information, *taken together*, are consistent or inconsistent with the hypothesis. That's where mathematical modeling can be helpful. You can use a mathematical model to combine all the bits of information and work out their implications.

The work of Meghan Duffy and Spencer Hall (2008) is a great example. They studied the population dynamics of the herbivorous zooplankton *Daphnia dentifera* in seven lakes in southwestern Michigan. *Daphnia* populations in these lakes experience epidemics of two parasites, the yeast *Metschnikowia bicuspidata* and the bacterium *Spirobacillus cienkowskii*. Both parasites are highly virulent: infection with either drastically reduces *Daphnia* survival and fecundity and also makes *Daphnia* more susceptible to fish predation. Yet, the two parasites are associated with very different host population dynamics: *Spirobacillus* epidemics are associated with substantial, rapid declines in *Daphnia* density; *Metschnikowia* epidemics are not. Why the difference? Duffy and Hall identified two possible explanations. First, *Metschnikowia*-infected *Daphnia* are subject to very selective predation by fish, meaning that infected individuals are much more likely to be consumed than are uninfected individuals. Fish predation on *Spirobacillus*-infected *Daphnia* is less selective. A parasite outbreak won't reduce host density if the infected hosts all get eaten before the outbreak can proceed too far. Second, *Daphnia* might rapidly evolve resistance to *Metschnikowia*, but not to *Spirobacillus*. Rapid evolution of resistance could terminate *Metschnikowia* outbreaks before they greatly reduce *Daphnia* density.

Before we go further, it's worth noting a few things about these two explanations. First, they're not mutually exclusive alternatives; both could be at work. Second, they might interact and maybe even cancel one another out. In particular, it's not clear if hosts could rapidly evolve resistance to *Metschnikowia* if they're all being eaten by selective predators. Third, whether either explanation or both actually work is a quantitative rather than a qualitative question. It's not enough to merely know whether or not fish selectively prey on infected hosts. You need to know if they're selective *enough* to terminate *Metschnikowia* epidemics before host density crashes, but *not* selective enough to terminate *Spirobacillus* epidemics before host density crashes. Similarly, it's not enough to know whether *Daphnia* have more genetic variation for *Metschnikowia* resistance than they do for *Spirobacillus* resistance. You have to know whether the difference is large *enough* to allow rapid resistance evolution to *Metschnikowia* but not *Spirobacillus*.

So Duffy and Hall built a mathematical model of host-parasite dynamics. They used data from laboratory assays, field observations, and other sources to estimate the model parameters, or in some cases realistic ranges of parameters that can't be estimated precisely. The model revealed that it's the combination of selective fish predation and rapid resistance evolution that terminates *Metschnikowia* epidemics before they can greatly reduce *Daphnia* density. The model also revealed additional, counterintuitive results—at least, results that seem counterintuitive until you do the math. Doing the math replaces your pre-mathematical intuitions with new, improved intuitions. For instance, if the overall fish predation rate weren't quite as high as it actually is, selective fish predation on *Metschnikowia*-infected hosts would actually lengthen epidemics, leading to bigger epidemic-induced reductions in host density. In this scenario, infection prevalence is very high, leading to longer, larger epidemics.

There's a joke in ecology that the answer to every ecological question is "It depends." Duffy and Hall (2008) is a nice illustration of how this needn't be an expression of ignorance. With the help of math, it can be an expression of knowledge: "It depends—and here's how, and why."

4. DIVIDE A WHOLE INTO PARTS

Duffy and Hall used math to figure out how various effects combine to generate a net effect. You can also use math to do the reverse: divide a whole into meaningful parts.

One common example in ecology is dividing gamma diversity (γ) into two components: alpha (α) and beta diversity (β; Whittaker 1960). For instance, imagine you have data on the abundances of each of many species, at each of many sites. You calculate some number that quantifies the diversity of the entire dataset—that's gamma diversity. Intuitively, gamma diversity depends on two components: diversity within sites and diversity among sites. The former component is alpha diversity, the latter component is beta diversity. Alpha and beta diversity might well depend on different factors, and/or depend on the same factors in different ways. For instance, alpha and beta diversity might depend in different ways on how organisms move among sites. Further, two study systems with the same gamma diversity might differ in alpha and beta diversity. So ecologists have derived ways to calculate gamma diversity and divide it into alpha and beta subcomponents.

Yes, *ways*, plural. There's no one agreed-upon definition of gamma diversity, and no one agreed-upon way to divide any given measure of gamma diversity into alpha and beta diversity (Lande 1996; Jost 2006; Jurasinski, Retzer, and Beierkuhnlein 2009; Tuomisto 2010; Baselga and Leprieur 2015; Ricotta 2017; Ricotta and Feoli 2024; see also chapter 7). Rather, different authors have proposed different partitions of gamma diversity, each of which has its own pluses and minuses. For instance, additive partitions define gamma, alpha, and beta diversity in such a way that $\gamma = \alpha + \beta$. Multiplicative partitions define gamma, alpha, and beta diversity in such a way that $\gamma = \alpha \times \beta$.

One can even go further and partition gamma diversity into more than two subcomponents. For instance, partitioning beta diversity into subcomponents associated with diversity at different nested spatial scales (e.g., sites within biogeographic regions within continents). As a general rule of thumb, always start with the most finely subdivided partition possible. You can always lump together some subcomponents if they turn out not to be informative. But if you never calculate those subcomponents in the first place, you won't know what you're missing.

Diversity partitions are perhaps the most familiar partitions in ecology, but they're far from the only ones. Modern coexistence theory (Chesson 2000; Ellner et al. 2019) is a partition: it divides a rare species' per capita growth rate into components attributable to different classes of coexistence mechanism (mechanisms that allow rare species to increase in abundance in the face of interspecific competition, rather than declining to extinction). The Price equation is a partition (Price 1970, 1972). George Price originally developed it to divide evolutionary change in the mean phenotype of a population into components attributable to natural selection (differential fitness of parents with different phenotypes) and biased transmission (any factor causing offspring phenotypes to deviate from those of their parents, on average). Price (1972) subsequently extended his partition to divide evolutionary change attributable to selection into subcomponents attributable to group-level selection versus individual-level selection. Many subsequent authors, including me, have extended and reinterpreted the Price equation to partition change over time in ecological variables (J. Fox 2006; J. Fox and Kerr 2012; Frank and Godsoe 2020), and to partition ecological versus evolutionary contributions to phenotypic change over time (Collins and Gardner 2009; Govaert, Pantel, and De Meester 2016). The Price equation has even been reinterpreted and extended to partition change in nonbiological variables, such as linguistic variables (B. Clark 2010).

One use of partitions is to prevent the researcher from overlooking possibilities. For instance, some collaborators and I used an extension of the Price equation to partition changes over time in average mammalian body size, as recorded in the fossil record at a site in Wyoming (Rankin et al. 2015). The Paleocene-Eocene Thermal Maximum (PETM) was a brief, intense global warming event approximately 55 mya. The average mammal at this site was smaller after the PETM than before. Why? Previous work identified two possible answers, not mutually exclusive: the mammal species at the site before the PETM might've shrunk on average, and the mammal species that colonized the site after the PETM might've been smaller on average than the species that were there already (Gingerich 2006; Secord et al. 2012). There's evidence for each of those possibilities. But previous work had overlooked a third possibility: species-level selection. That is, species present before the PETM might've varied in their propensity to speciate and/or go extinct in a way that depended on body size. All else being equal, mean mammalian body size will decline if smaller-bodied species speciate more, and/or go extinct less, than larger-bodied species. My collaborators and I (Rankin et al. 2015) used the Price equation to show that those three possibilities are in fact the only three, and calculated their values from fossil data. It turns out that species-level selection during the PETM actually favored larger-bodied species, partially canceling out reductions in body size due to both shrinkage of persisting species and post-PETM colonization of the site by small-bodied species unrelated to the residents.

It's interesting that debates about partitions tend to run to type. For instance, there's no one unique way to partition evolutionary change in mean phenotype into selection versus transmission, or to partition evolutionary change due to selection into individual versus group selection (Okasha 2004). That's because we have conflicting intuitions about what "selection" and "transmission" mean, and what "individual selection" and "group selection" mean. The same is true for debates over how to partition gamma diversity into alpha and beta diversity: we have conflicting intuitions about what "alpha" and "beta" diversity mean. Doing the math may not reveal which of those intuitions should be discarded, if any. But doing the math can at least make you aware of the conflict.

5. COMPARE AND CONTRAST DIFFERENT MODELS OF THE SAME THING

Comparing the predictions of different models to one another, and to data, allows us to apportion responsibility for any errors in those predictions. Which features of the world actually explain the phenomenon of interest, and which don't matter? A result that holds for all models presumably reflects one or more shared features of those models, and presumably is independent of model-specific assumptions. It's what Richard Levins (1966) famously called a "robust theorem" (see chapter

7 for discussion). In contrast, if a result holds for some models but not others, comparing among the models allows identification of the assumptions or conditions on which the result depends.

G. Harrison (1995) provides an example of learning from the contrasts between different models. Gary Harrison's goal was to explain the results of a classic series of experiments by Leo Luckinbill (1973) on the population cycles of the protist predator *Didinium nasutum* and its protist prey species *Paramecium aurelia*. In laboratory culture, these two species exhibit an extinction-prone predator-prey cycle, with predators invariably going extinct before their prey. Luckinbill found that, by lowering the resource enrichment (carbon and nutrient concentration) in the culture medium, he could reduce the maximum abundance and maximum growth rate of the bacteria on which *Paramecium* feeds. This change reduced the amplitude of the predator-prey cycles, allowing predator and prey to persist for multiple cycle periods before either species went extinct. If lower resource enrichment was combined with thickening the culture medium (thereby reducing the movement rates of both predator and prey species, and so reducing their per capita encounter rate), cycle amplitude was reduced still further, and the cycles persisted indefinitely.

Harrison used a series of increasingly complex predator-prey models to explain these results. The simplest model, the Rosenzweig and MacArthur (1963) model, considers a prey population that grows logistically in the absence of predators, and a predator with a type II functional response, a constant birth rate per prey consumed, and a constant rate per capita mortality rate independent of prey or predator density. When fitted to the time series data in order to estimate its parameters, this model qualitatively captured Luckinbill's results. The modeled predator and prey exhibited population cycles, which decreased in amplitude when enrichment was reduced (modeled as reduced prey carrying capacity), and the predator-prey encounter rate was reduced (modeled as reduced predator attack rate). But the Rosenzweig–MacArthur model produced a poor quantitative fit to Luckinbill's data. It produced cycles of the wrong period, the cycle peaks were much too sharp, and the cycle nadirs too long-lasting. Harrison tried several more complex alternative models that made more biologically realistic assumptions than the Rosenzweig–MacArthur model. Most of them also produced poor fits, or produced good fits only with unrealistic parameter values. Some of them failed to even qualitatively reproduce the effects of reduced enrichment and medium thickening. However, a model that modified the Rosenzweig–MacArthur model with an asymmetrical type III predator functional response and time-lagged predator numerical response gave an excellent quantitative fit, and accurately predicted the effects of reduced enrichment and medium thickening. Reassuringly, these modifications have biological interpretations. The predator *Didinium* stops hunting when prey become sufficiently rare (pers. obs.), a behavior that would generate an asymmetrical type III functional response. And there is indeed a time lag between when *Didinium* consumes a *Paramecium* and when it reproduces. However, even this model continued to make some systematic errors, predicting prey densities that dropped too low and failing to reproduce the observation that *Didinium* usually goes extinct before *Paramecium*. This suggests that even the best-fitting model still omits or misdescribes one or more features of this predator-prey interaction. The best-fitting model also makes new predictions that could be tested in future experiments.

As an example of learning from a shared feature of different models, consider models of food-web topology. Food-web topology refers to the arrangement of predator-prey interactions linking together consumers and consumed. We can measure many different aspects of food-web topology: the mean length of all food chains connecting a species at the base of the food web to a top predator; the proportion of species that are cannibalistic; the frequency distribution of number of prey species per predator species; and so on. A good model of food-web topology will accurately reproduce the metrics observed in real food webs. Williams and Martinez (2000) proposes a model to explain many metrics of food-web structure. The model takes as inputs the number of species, and the number of predator-prey links, then links up species at random subject to the constraint that species form a "hierarchy": species can feed only on those that rank lower than themselves in the hierarchy, but not too much lower. The link-assignment algorithm is actually a bit more complicated than I just described, but that's the gist. Other researchers have proposed other models that take the same input parameters, but that assign links to species using different link-assignment algorithms (Cattin et al. 2004;

J. Cohen and Newman 1985). The differences among those algorithms turn out to be of minor importance, however. It turns out that any link-assignment algorithm will work reasonably well, so long as it assumes or produces a hierarchy of species, such that higher-ranked species tend to consume those of lower but not too much lower rank (Stouffer et al. 2005). Explaining that hierarchy is therefore an important task for empirical research. Explain that, and you've pretty much explained why food webs have the topologies they do. In some food webs, the hierarchy likely reflects predators consuming prey smaller, but not too much smaller, than themselves (P. Warren and Lawton 1987; J. Cohen, Jonsson, and Carpenter 2003). But in other food webs, body size can't explain the hierarchy—think of insect herbivores and the terrestrial plants they consume.

Here's a third example, from my own work. Jacob Rovere and I (Rovere and Fox 2019) showed that rare lake zooplankton species (those that have low relative abundance on average) tend to bounce back very strongly when their relative abundances drop extremely low. A small drop in the relative abundance of a rare species leads to a large increase in its expected per capita growth rate. In contrast, common zooplankton species bounce back much less strongly. Their per capita growth rates increase only modestly when their relative abundances drop. It's tempting to explain the tendency of rare species to bounce back strongly following declines in their relative abundances by appealing to the "no $10 bills lying on the ground" argument, after the old joke about economists (google it if you don't know it). You never see $10 bills lying on the ground, because if any fell on the ground they'd be picked up very quickly. Analogously, any rare species that doesn't bounce back strongly following a decline in its relative abundance would go extinct rapidly, and so wouldn't be observed because it would no longer exist. But that tempting explanation is unjustified. It turns out that rare species tend to bounce back more strongly than common species in almost any ecological model one can construct, even those in which species can't go extinct. As long as there's some finite upper limit to how fast any species can increase in abundance, and as long as each species has some relative abundance at which its birth and death rates are equal, rare species are likely to bounce back more strongly than common ones (Rovere and Fox 2019; fig. 5.5).

Figure 5.5. Rare species will tend to bounce back more strongly from near-zero density than will common species

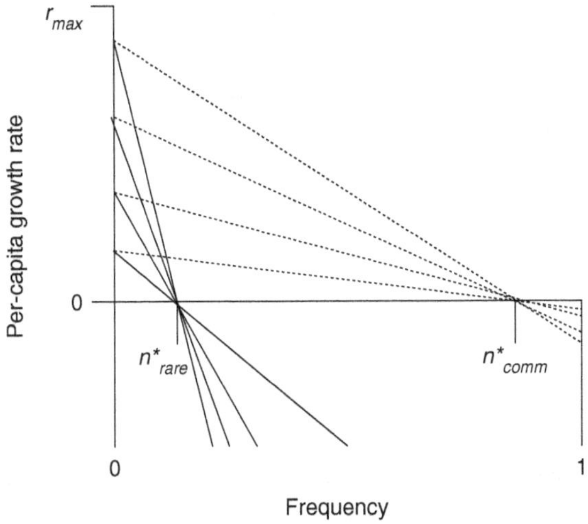

Graph of per capita growth rate versus frequency (relative abundance) of two species, one common and the other rare. Each species has some equilibrium frequency at which its per capita growth rate equals zero. For each species, negatively sloped lines illustrate some of the many possible relationships between that species' per capita growth rate and its frequency. The lines associated with the rare species tend to have steeper negative slopes than those associated with the common species. This implies that, if the density of rare species were to decline in frequency a bit, its per capita growth rate likely would increase by more than would that of the common species if it were to decline in frequency a bit.

Source: Rovere and Fox (2019).

6. USE A MATHEMATICAL MODEL AS A "NULL MODEL"

That is, use a mathematical model to capture larger or more obvious effects, but deliberately omit some hypothesized smaller, subtler, or more interesting effect. When compared to data, the model serves as a "null model" that factors out or controls for the larger or more obvious effects. This allows the smaller, subtler, or more interesting effect to be detected, and perhaps even quantified.

As discussed in chapter 4, ecologists' attempts to use mathematical models in this way have a pretty poor track record. I won't repeat that discussion here.

One successful example of this use of mathematical models in ecology is to subtract out "sampling effects" from the results of BEF experiments. Recall from chapter 2 that many ecological experiments, starting with the first one ever conducted, have asked whether a diverse mixture of many plant species will produce more plant biomass per unit area than would a less diverse community of just one or a few plant species (Hector and Hooper 2002). BEF experiments ordinarily employ a "random draws" design. The experimenter identifies a "pool" of *S* species to be included in the experiment. The experiment includes single-species plots of each of the *S* species, termed monocultures, and maximally diverse plots comprising all *S* species growing together. The experiment also includes several randomly selected mixtures of species for each of the other diversity levels included in the experiment. For instance, an experiment with $S = 16$ species might include several two-species plots, each with two randomly chosen species from the pool, several randomly chosen four-species mixtures, and so on. Total density is held constant across species richness levels to avoid confounding species richness and total density.

A concern about this design is the potential for a "sampling effect": more diverse plots have greater odds of including a very productive species, and so might tend to outproduce less diverse plots for this arguably trivial reason (Huston 1997; Aarssen 1997). Michel Loreau and Andy Hector (2001) showed how to use data from the monocultures to calculate the expected biomass of each mixture plot, assuming that species' per capita productivities are independent of which and how many other species are in the plot. If species' per capita productivities are constant, then the sampling effect is the only determinant of mixture biomass. That's our null model. It turns out that observed mixture biomasses usually differ substantially from the values expected under this null model.

Of course, sometimes a null model that captures the larger or more obvious effects reveals that there aren't any smaller or more subtle effects to discover. The work of Xiao Xiao, Kenneth Locey, and Ethan White (2015) provides a nice example. The researchers set out to explain a widespread ecological pattern known as Taylor's law. Taylor's law states that if the abundance of a species is observed at many sites or times, the variance in abundance among sites or times will be proportional to the average abundance raised to some power b (Taylor 1961). Empirical work on many species shows that b varies within a limited range; it's almost always between 1 and 2 (Taylor and Woiwod 1982; Xiao, Locey, and White 2015). Why should the variance in abundance be proportional to mean abundance raised to a power between 1 and 2? Various ecological explanations have been proposed, all of which have drawbacks and limitations (reviewed in Xiao, Locey, and White 2015). For instance, explanations based on how individual organisms move in space (e.g., Hanski 1982) can't explain why Taylor's law holds for temporal as well as spatial variation in abundance. Xiao, Locey, and White proposed a simple null model: divide the total number of individual organisms N randomly across all the sites or times included in the dataset (including those sites or times at which no individuals were observed). More-abundant species are those with higher N. "Randomly" means that any given observation of an individual is equally likely to have occurred at any site or time in the dataset. (I'm glossing over some technical details here, but that's the gist.) Obviously, there are many possible ways to randomly divide N observations among x sites or times, some more probable than others. It turns out that the vast majority of possible divisions of observations among sites or times lead to Taylor's law: variance in abundance among sites or times is proportional to mean abundance raised to the power $1 < b < 2$ (fig. 5.6). So the general form of Taylor's law—it's a power law with an exponent between 1 and 2—doesn't require an explanation in terms of any ecological factors omitted from the null model in Xiao, Locey, and White (2015).

Sometimes—rarely, but sometimes—it's possible to push null models even further. Instead of using a null model that omits a factor or effect merely to detect that factor or effect, we can use the model to also estimate the quantitative magnitude of the omitted factor or effect. Consider Ellner et al. (2019) on quantifying the "storage effect," a mechanism by which spatial or temporal variation in environmental conditions can promote the coexistence of competing species. A storage effect is defined by certain statistical correlations between ecological variables that vary over space and time, such as environmental conditions and species' per capita growth rates. Ellner et al. (2019) shows how to estimate the strength of the storage effect by calculating a measure of how strongly species coexist in the absence of the storage effect. The

Figure 5.6. A null model for Taylor's law

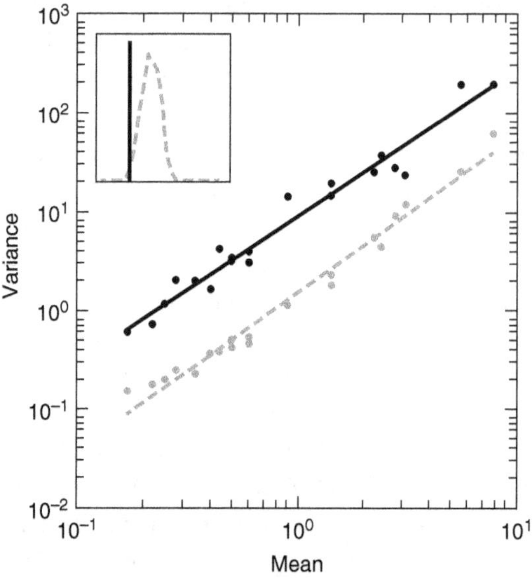

Log-log plot of an observed relationship between the mean and variance of abundance (*black points and solid black regression line*), and the average mean-variance relationship from a null model that randomly shuffles the observations among sites (*gray points and dashed gray regression line*). The slope of the regression line estimates *b*, the value of the exponent in Taylor's law. Inset gives estimated *b* for the observed data (*vertical solid line*) and the distribution of *b* values generated by the null model (*dashed line*). The observed mean-variance relationship and the mean-variance relationships generated by the null model both are well described by linear regressions with slopes *b* between 1 and 2, although the mean *b* from the null model doesn't equal the estimated *b* for the observed data.

Source: Modified from Xiao, Locey, and White (2015).

difference between (i) how strongly species coexist and (ii) how strongly they would coexist in the absence of the storage effect, estimates the strength of the storage effect.

7. USE TWO DIFFERENT MATHEMATICAL MODELS TO BRACKET REALITY

That is, use two different mathematical models to define the extremes of a continuum of cases, between which reality is presumed to fall. You do this because reality is too complicated to model, but the extremes that bracket reality are simpler.

This use of mathematical models is rare in ecology, for good reason: ecological reality often doesn't fall on a continuum between different limiting cases, thanks to nonlinearities, time lags between cause and effect, feedback loops, and other complications.

For instance, consider simple metapopulation models describing the dynamics of spatially separated populations connected by random dispersal of organisms among populations (Yaari et al. 2012). In the limiting case of no dispersal, the populations are disconnected and independent of one another. Extinct populations are never recolonized, and so the metapopulation persists only as long as the longest-lived population persists. In the limiting case of very high dispersal, the metapopulation is effectively one large population. Organisms are constantly moving randomly among populations, like molecules moving around a well-stirred reactor vessel in a chemistry laboratory. This rapid dispersal homogenizes the populations, the abundances of which all fluctuate in synchronized fashion—they all increase in abundance, and decrease, at the same time. The metapopulation persists until such time as all the populations crash to extinction simultaneously. Metapopulations with high dispersal tend to last somewhat longer than otherwise equivalent metapopulations with no dispersal, and so we might naively expect that metapopulations with intermediate dispersal rates would persist for an intermediate length of time. But in fact, intermediate dispersal rates increase metapopulation persistence time far beyond what is possible with either no or high dispersal. Intermediate rates of dispersal allow rapid recolonization of extinct populations without synchronizing the abundance fluctuations of different populations.

As far as I'm aware, no ecologist has ever used metapopulation models with zero or high dispersal to come to a mistaken understanding about metapopulations with intermediate dispersal. But there are cases in which undue focus on limiting cases has proven to be a misleading guide to a more complicated ecological reality that's not intermediate between the limiting cases. The authors of Leibold et al. (2004) reviewed the theoretical literature on metacommunity models—models of spatially separated ecological communities linked by together by organisms dispersing from one community to another. They classified metacommunity models into four types, each focused on a different limiting case. For instance, "neutral" models consider the limiting case in which environmental conditions are the same everywhere and all competing species are identical to one another. In neutral models, variation in species abundances within and among communities is driven purely by dispersal, plus random fluctuations in species' birth and death rates (demographic and environmental stochasticity). In contrast, "species sorting" models consider the limiting case in which ecological differences among species and locations are so large as to completely dictate species' abundances within any given local community, with dispersal and demographic stochasticity having no detectable effect. The historical review in Leibold et al. (2004) was intended as a road map of the theoretical literature existing at the time, not as an exhaustive classification of all possible metacommunity models, or even of all possible limiting cases (see also B. Brown et al. 2017). But others interpreted Leibold et al. (2004) as defining the space of possibilities, or at least as defining all possible limiting cases, and proposed empirical methods to classify natural metacommunities into the four types that the researchers identified (Cottenie 2005; Soininen 2014). Those methods turned out to fail badly when applied to simulated data generated from metacommunity models structured by known processes (Gilbert and Bennett 2010; Logue et al. 2011; B. Brown et al. 2017; Barbier, Bunin, and Leibold 2025). One important reason for those failures is that metacommunities structured by a mix of different processes do not exhibit behavior intermediate to that of the four limiting cases reviewed by Leibold et al. (2004).

Klausmeier et al. (2004) provides a more successful example of this use of mathematical models in ecology. The researchers modeled how unicellular marine phytoplankton might optimize their acquisition of three key resources they need to grow: light, nitrogen (N), and phosphorus (P). Algae use chloroplasts to acquire light, nitrate uptake proteins on the cell surface to acquire nitrogen, and so on. Light-limited algae should make more chloroplasts, at the cost of making fewer structures to acquire other non-limiting resources, nitrogen-limited algae should make more nitrogen-acquisition structures, and so on. Because different cellular structures are made of different ratios of N:P, algae limited by different resources have different optimal N:P ratios, ranging from 8.2 to 45.0 according to the data that the researchers used to parameterize their model. However, limitation of algal growth by any single resource is a limiting case that often only prevails temporarily, due to spatial and temporal variation in resource availability in the ocean. The authors of Klausmeier et al. (2004) therefore argued that real unicellular marine phytoplankton will vary in their N:P ratios, in between the N:P ratios that would be optimal under growth limitation by different single resources. This informal heuristic argument might well be wrong! Ideally, we'd actually model the intermediate cases. But absent an explicit model of the intermediate cases, there is some reason to think that the intermediate cases do fall on a continuum between the limiting cases: the N:P ratios of unicellular marine phytoplankton mostly do fall within the limits predicted by the model in Klausmeier et al. (2004; fig. 5.7). That obviously doesn't show that the model is correct; there could be other explanations for observed variation in marine phytoplankton N:P ratios. But the data are at least consistent with the model, and so further research might be warranted.

Figure 5.7. Interspecific variation in phytoplankton N:P ratios may reflect optimization for different environmental conditions

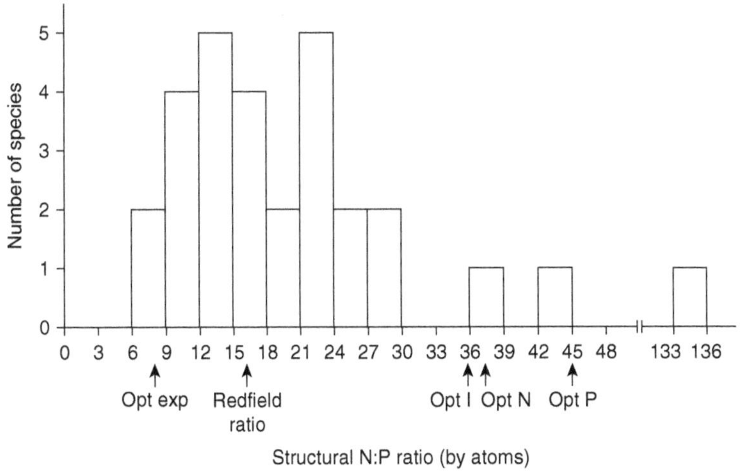

Bars are a histogram of N:P ratios of 29 species of phytoplankton. The average N:P ratio of phytoplankton, 16:1, is known as the Redfield ratio. Also marked are theoretically predicted optimal (fitness-maximizing) N:P ratios for phytoplankton growing exponentially in nutrient- and light-rich water ("Opt exp"), and for phytoplankton whose growth is limited by scarcity of light ("Opt I"), N ("Opt N"), or P ("Opt P"). The theoretical predictions define four different limiting cases, which bracket most of the observed variation in N:P ratios.

Source: Modified from Klausmeier et al. (2004).

Conclusions

It's not just plant species that perform different ecosystem functions. Any set of species you care to name performs various ecosystem functions. Analogously, it's not just mathematical models that can be used to do many different jobs. What's true of mathematical modeling is true of other approaches to ecological research. For instance, there are various different sorts of experiments that can be conducted for various different purposes (Wootton and Pfister 1998; J. Fox and Lenski 2015). The same is true of observational studies, studies of artificial systems like microcosms, and any other research approach. By showing how blanket criticisms of mathematical modeling in ecology overlook its many uses, I hope to awaken the reader to the many uses of other research approaches.

6 *Fighting Lack of Diversity: The Value of Contrarians*

So far, this book has been all about harnessing the diversity of scientific ideas, goals, and approaches in science, drawing on an analogy to BEF experiments. What does it take for a diverse plot to outperform a monoculture? Analogously, what can ecologists learn from BEF experiments, in order to get the most out of our diverse ideas, goals, and approaches?

But now we've reached the limits of the BEF analogy. That analogy can't teach us everything we need to know about harnessing diversity of goals, ideas, and approaches in science. So in the remaining chapters, we're going to set aside this analogy.

For instance, here's a question that comes up in scientific research, but not in BEF experiments: what if you don't have diversity in the first place? How do you get some? What if you don't have diverse scientific ideas, but instead have the scientific equivalent of a monoculture?

Well, usually that's a good thing! Usually, it means that science has converged on the truth, so that no further research is needed. Geologists all agree on how old Earth is. Evolutionary biologists all agree that humans evolved from an ape-like ancestor. Climatologists all agree that anthropogenic climate change is real. And so on. The only people who think we need to increase the diversity of ideas about those topics are creationists and climate change denialists.

But a scientific monoculture isn't *always* a good thing. It doesn't *always* mean that science has converged on the truth. Sometimes it just means that science has converged. Sometimes the scientific consensus is dubious or wrong, for whatever reason (and it could be for various reasons, most of which can happen even though almost all scientists are competent and ethical). I emphasize that I *don't* think it's *common* for the scientific consensus to be dubious or wrong. But when it is, that's very bad.

Further, it's not bad just because it's bad for everyone to believe something that's dubious or wrong. It's also bad because it's hard to fix. When the scientific consensus is dubious or wrong, the normal error-correcting methods of science tend to fail, just when they're needed most. Imagine some scientific claim that scientists all think is well established, even though it's actually not. Peer reviewers who think—incorrectly—that the claim is well established will tend to dismiss out of hand manuscripts questioning it. Students will be taught the claim, thereby perpetuating belief in it. No scientists will want to double-check whether the claim is true or not. Why would you waste your time double-checking a claim that everyone thinks is well established? And so on. The very same mechanisms that nudge the scientific consensus toward the truth also help maintain the consensus when the consensus is dubious or false.

So how does a scientific field break out of a "bad equilibrium" in which the consensus is dubious or wrong? There's really only one option: you need a contrarian. Somebody who's prepared to push back against the consensus. Provide some diversity of thought, so that selection effects can start to operate and push the field toward a better equilibrium.

Easier said than done, of course. But definitely doable. In this chapter, we'll look at several case studies of contrarians in ecology (and in one case, paleontology), all of whom successfully injected some diversity of thought into a dubious or incorrect scientific consensus.

Maybe *Anolis* Lizards *Aren't* Territorial

Anolis lizards, especially those found on Caribbean islands, have long been a popular model system in evolutionary ecology—for instance, for questions about how competition influences evolution (Roughgarden 1995; Losos 2009).

The basics of anole natural history are well known because anoles have been studied intensively for a long time. In particular, anoles exhibit territorial polygyny. A territory-holding male defends his territory against incursions by conspecific males, thereby gaining exclusive mating access to the females who live in smaller territories within the male's larger territory (Losos 2009).

Except that can't possibly be "well known," because it can't possibly be true. We know it can't be true because molecular methods show that many female anoles produce offspring sired by multiple males (Johnson et al. 2021 and references therein). Which raises the question of how so many ecologists studying anoles could've gotten it wrong for so long.

That's a very good question, and it has a very good answer (Kamath and Losos 2017, 2018, which I'm summarizing). Briefly, the first studies of territoriality in anoles didn't last long enough, and didn't have large enough sample sizes, to really establish both key ingredients of territorial polygyny: site fidelity and exclusivity. Unless anoles really do spend their *entire* reproductive lives in the same locations (site fidelity), and unless males really do *completely* exclude other males from their territories (exclusivity), females will have opportunities to mate with multiple males. That's very difficult to establish, and it's not what early studies established. Rather, early studies—many conducted in laboratory environments rather than in nature—reported that male anoles are aggressive toward conspecific males. Many male anoles certainly *try* to defend exclusive territories against other males; but whether *all* males try to do so, and do so *successfully*, is another matter. Other early studies inferred site fidelity from re-observing the same individuals in the same locations a few times over the course of a few weeks (a period of time much shorter than a single mating season). Cases in which lizards weren't re-observed in the same location, and cases in which female territories overlapped those of multiple males, were noted but not emphasized by study authors. Subsequent studies of territorial and mating behavior in anoles took it as given that anoles exhibit territorial polygyny, and so weren't designed to check whether anoles do in fact exhibit territorial polygyny. Indeed, subsequent studies often were designed so as to prevent observation of anything but territorial polygyny—for instance, by excluding from the dataset any cases of multiple males occupying the same site.

Many things that ecologists "know" are like this. More science than we care to admit is a bit like a game of telephone. We start with some relevant data, or perhaps a classic case study, from which some broader conclusion is drawn. That broader conclusion gets repeated, and then the repetition gets repeated—in textbooks, in seminars, in passing citations, and likely getting oversimplified or garbled along the way. Heck, it's surprisingly common for ecology papers to be cited in support of claims they didn't even make (Todd, Yeo, et al. 2007; Todd, Guest, et al. 2010; D. Drake et al. 2013), and for ecologists to cite sources that don't even exist (Šigut et al. 2017)! We also all "know" that many subsequent studies "support" some broad conclusion. Except that all those subsequent studies really do is build on that broad conclusion, by assuming it's true. And so we all end up "knowing" a lot of things that aren't true, or at least aren't as well established as we think they are. We already saw another example in chapter 4: everybody "knows" that habitat fragmentation usually reduces biodiversity—even though it usually doesn't. Everybody "knows" that 8% of Earth is covered by lichens—even though that claim was originally based on no data whatsoever (Drotos, Larson, and McMullin 2024). Everybody "knows" that bacterial cells in and on the human body outnumber human cells by 10 to 1—even though our best estimate of the true ratio is 1 to 1 (Sender, Fuchs, and Milo 2016). Everybody "knows" that 80% of the world's biodiversity is found on Indigenous territories—even though that's a garbled misinterpretation of a very different claim (Fernández-Llamazares et al. 2024). And so on.

Maybe Ocean Acidification *Doesn't* Disrupt Fish Behavior

One consequence of increasing atmospheric CO_2 is ocean acidification. Increasing partial pressure of CO_2 in the atmosphere increases the concentration of dissolved CO_2 in water, reducing ocean pH. In a high-profile series of papers, Danielle Dixson, then a PhD student in the lab of marine ecologist Philip Munday, reported severe negative effects of ocean acidification on reef fish behavior (Dixson, Munday, and Jones 2010; Dixson, Abrego, and Hay 2014; Munday, Dixson, Donelson, et al. 2009; Munday, Dixson, McCormick, et al. 2010). For instance, in laboratory assays, reef fish placed in acidified water chose to spend almost all their time in water containing the olfactory cues of predatory fish. In contrast, reef fish in the control treatment avoided the olfactory cues of predators. In other laboratory assays, ocean acidification disrupted the homing abilities of larval reef fish, suggesting that ocean acidification will prevent fish from choosing suitable habitats in which to grow into adults. These results were published in high-impact journals and prompted hundreds of follow-up studies looking for effects of ocean acidification on the behavior of other species (Clements et al. 2022). The scientific consensus, as stated in IPCC reports, was that ocean acidification disrupts normal reef fish behavior.

Fortunately, that consensus wasn't quite universal (Browman 2017; T. Clark et al. 2020). One reason for doubt was the lack of a physiological mechanism by which ocean acidification might affect reef fish behavior. Reef fish can maintain their own tissue pH within the normal physiological range, even in water much more acidic than used in any behavioral assay. Further, reef fish in nature experience natural, non-anthropogenic fluctuations in ocean water pH without any apparent disruption of their normal behavior. In light of our background knowledge of fish physiology, it's quite surprising that ocean acidification would affect fish behavior at all. But perhaps that just meant that some unknown physiological mechanism was waiting to be discovered.

Then the authors of T. Clark et al. (2020) raised a second reason for doubt. They tried to replicate Danielle Dixson's original results—and failed. In their experiments, there wasn't even a hint of an effect of ocean acidification on fish behavior. The authors also pointed out that Dixson's original results reported implausibly large differences between treatment and control means, and implausibly small within-treatment standard deviations. Even under carefully controlled laboratory conditions, replicate fish don't ordinarily all behave *exactly* the same way when subjected to the same experimental treatment, as Dixson's fish purportedly did. More broadly, the results of Dixson, and other Munday lab members, stuck out like a sore thumb in the context of all published results on ocean acidification and fish behavior (Clements et al. 2022; fig. 6.1). Meta-analysis revealed a massive "decline effect" in studies of ocean acidification on fish behavior—the estimated mean effect size declined rapidly in magnitude over time, as more and more studies were published (Clements et al. 2022; fig. 6.1). But if you dropped results from the Munday lab, the decline effect vanished. That is, the other labs that had followed up the Munday lab's early work consistently found weak effects (at best) of ocean acidification on fish behavior (Clements et al. 2022; fig. 6.1). In summary: from a statistical perspective, the original results from Dixson's work looked almost too good to be true.

Figure 6.1. Decline in the estimated effect of ocean acidification on reef fish behavior

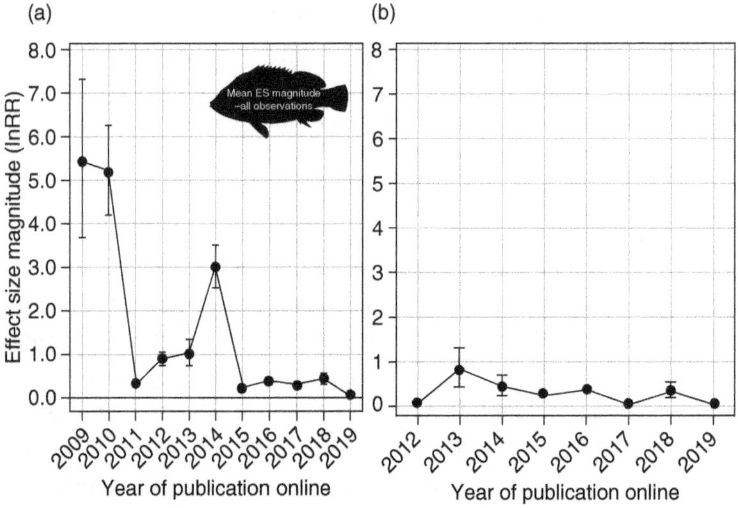

The estimated mean effect declined rapidly over time (a), because early studies finding large effects all came from one lab. Removing that lab's work from the dataset eliminates the decline effect (b). We now know that early studies finding large effects were fabricated.

Source: Modified from Clements et al. (2022).

That's because they weren't true. They were fake. A 2021 investigation by news reporters from the journal *Science*, and a subsequent 2022 investigation by the University of Delaware, Dixson's employer, revealed that Dixson had fabricated her results (Enserink 2021, 2022). She never conducted the experiments she claimed to have conducted. Instead, she created her data files by copying and pasting numbers that would generate the desired (and implausibly huge) effects of ocean acidification. The University of Delaware fired her from her tenure-track faculty position.

Which in retrospect makes a few ecologists' real-time reactions to the work in T. Clark et al. (2020) and Clements et al. (2022) a little . . . let's go with "striking." The authors of T. Clark et al. (2020) and Clements et al. (2022) were publicly accused by both Munday as well as some other marine ecologists unconnected with the Munday lab of being bad scientists. They were accused of seeking to advance their own careers by tearing down the work of others rather than doing their own research (Enserink 2021, 2022).

Maybe Local Biodiversity *Isn't* Declining

BEF experiments ask what will happen to ecosystem function if some species are lost. Further, those experiments invariably ask about local-scale biodiversity declines. They ask about loss of species from a patch of grassland a few square meters in area, or from a small pond, or from some other small patch of habitat. That raises the question of whether or how the results of BEF experiments scale up to entire landscapes or continents (Gonzalez, Germain, et al. 2020), but leave that aside. For now, the point is just that those experiments ask about the local-scale effects of local-scale species loss.

But *are* species being lost at local scales? Is it in fact the case that a typical patch of grassland, say, or a typical pond, contains fewer species than it did decades ago? If you read my BEF papers, and many other BEF papers, you could be forgiven for thinking that the answer to that last question is "Yes, obviously." My BEF papers, like pretty much everyone else's, all start off by saying that species are going extinct worldwide due to human activities, thereby raising concern about the consequences of species loss for ecosystem function. It's standard boilerplate that reflects a consensus among ecologists: biodiversity is declining everywhere, and that's bad.

But that boilerplate glosses over a bunch of complications (Maier 2012)—in particular, issues of spatial scale. Just because the planet as a whole has lost numerous species due to human activities since the end of the Pleistocene (R. Lemoine, Buitenwerf, and Svenning 2023) doesn't mean that a typical local-scale patch of habitat has lost species. After all, human activities also allow some species to colonize habitat patches they otherwise wouldn't be able to survive and reproduce in, or even reach in the first place (e.g., Seebens et al. 2016). So maybe a typical patch of habitat these days harbors *different* species than it used to, but not *fewer*.

Turns out that's exactly what's going on. The authors of Vellend, Baeten, et al. (2013) and Dornelas et al. (2014) compiled all the data they could find from long-term monitoring of species richness in local-scale patches of habitat. Vellend, Baeten, et al. (2013) considers local-scale terrestrial plant species richness; Dornelas et al. (2014) considers animal and plant species richness in terrestrial, marine, and freshwater habitats. Both groups found the same thing: on average, local species

Figure 6.2. Local plant species richness is not consistently declining (or increasing)

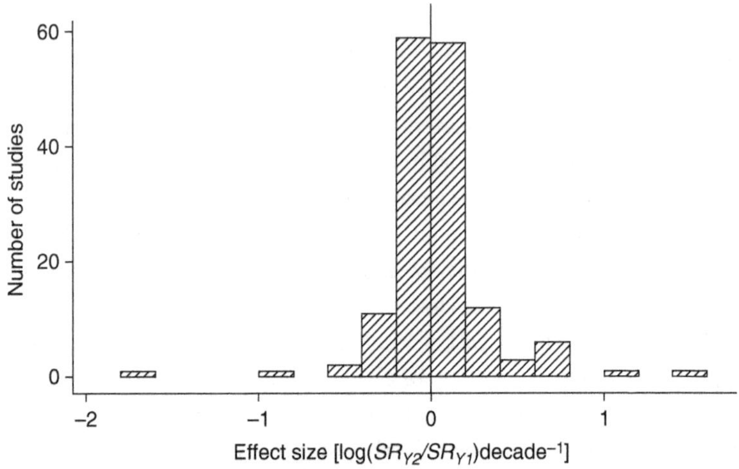

Histogram of the log-transformed ratio of final plant species richness (SR_{Y2}) to initial plant species richness (SR_{Y1}), expressed as a per decade rate of change, in long-term monitoring plots around the world. A log-transformed ratio of 0 indicates no change in species richness, negative ratios indicate declining richness, positive ratios indicate increasing richness.

Source: Modified from Vellend, Baeten, et al. (2013).

richness hasn't declined (or increased) over time (fig. 6.2). Rather, there's been extensive turnover of species composition—replacement of some species by others. The net result of which sometimes is a change in local species richness (typically a small change). But increases and decreases in richness are equal in frequency and magnitude.

Much to my surprise, some prominent ecologists didn't believe these results, subjecting them to various unfounded technical criticisms (Cardinale 2014; Gonzalez, Cardinale, et al. 2016; Vellend, Dornelas, et al. 2017). At least one ecologist publicly questioned the professional integrity of the authors of Vellend, Baeten, et al. (2013), accusing them of deliberately downplaying the biodiversity crisis in order to attract publicity and further their own careers (see comments on McGill 2014). Fortunately, those criticisms aren't worth dwelling on, because to everyone's credit they're now moot. Several of the authors, and several of their critics, subsequently began collaborating to compile additional data and conduct new analyses. These collaborative analyses recovered the conclusions in Vellend, Baeten, et al. (2013) and Dornelas et al. (2014), while revealing additional nuances. In particular, marine localities tend to exhibit more-rapid compositional turnover than do terrestrial or freshwater localities (Blowes et al. 2019).

I'll be curious to see how long it takes to develop a new consensus that local species richness is not declining on average. There's a huge "first mover" advantage in science. Widely cited claims typically continue to be widely cited even many years after they've been refuted (Tatsioni, Bonitsis, and Ioannidis 2007). Even when science has a diversity of competing claims to select from, as ecology now does with respect to the prevalence of local biodiversity declines, the selection process often proceeds very slowly (decades or more).

Maybe Ediacaran Organisms *Weren't* Built on the Same Principles as Modern Organisms

Our previous cases all involved overturning a consensus based on shaky foundations (or fake foundations, in the case of the effects of ocean acidification on fish behavior). But scientific contrarians can have value even when the consensus they're pushing back against is reasonably well founded, and even when the pushback itself turns out not to be entirely well founded. Pushing back against the prevailing consensus by proposing alternatives to it, and exposing gaps in it, can spur new thinking and new research. Even if the pushback doesn't pan out, the prevailing consensus ends up on firmer foundations than it otherwise would've had.

The best example I know of comes from paleontology rather than ecology. Consider Ediacaran organisms: the soft-bodied organisms dating from just before the Cambrian explosion, approximately 530 mya (fig. 6.3; this example comes from Max Dresow [2023, 2024a, 2024b], whose work I'm summarizing and paraphrasing). Ediacaran organisms don't exhibit any obvious resemblance to any extant species. This makes them difficult to place on the tree of life, and makes it difficult to infer how they made a living. Nevertheless, as of the late 20th century, the scientific consensus reflected the pioneering work of Martin Glaessner. Glaessner classified most Ediacaran organisms into extant phyla, based

Figure 6.3. *Parvancorina minchami,* an Ediacaran fossil

Source: Courtesy of Matteo de Stefano/MUSE.

on structural features putatively shared with extant phyla. For instance, Glaessner classified *Parvancorina* as an arthropod (fig. 6.3).

Adolf Seilacher (1989) questioned that consensus. Seilacher argued that Glaessner had been too quick to spot putative shared structures of Ediacaran and extant organisms. Instead, he should've placed more weight on "functional" and "fabricational" considerations. Basically, Seilacher (1989) argued that we shouldn't focus on whether an Ediacaran organism superficially resembles this or that extant organism. Rather, we should focus on how Ediacaran organisms were built and how their bodies worked. Seilacher (1989) inferred that many Ediacaran organisms were compounded from a shared structural element called a "pneu structure"—more or less a jelly-filled tube. These pneu structures were quilted together as mats on the ocean floor. They lacked mouths and anuses, so Seilacher speculated that they harbored symbiotic microbes that could extract energy from seawater. He also speculated that they were sitting ducks for the mobile predators that evolved during the Early Cambrian, thus explaining why they went extinct at that time. Seilacher

(1989) didn't classify Ediacaran organisms on the tree of life, but a few years later, Seilacher (1992) classified them into a new *kingdom* of life (!).

Other paleontologists quickly poked huge holes in this wild speculation. For instance, many Ediacaran organisms had distinct front and back ends; Seilacher was simply incorrect to say they didn't. But despite the criticisms, Seilacher's speculative hypothesis also attracted a lot of positive interest. In particular, Seilacher (1989) spurred other paleontologists to reconsider the classification of Ediacaran organisms with fresh eyes and new analyses. Discussing, and in some cases trying to refute, Seilacher's claims prompted paleontologists to realize they didn't really know as much about Ediacaran organisms as they had thought. They also realized there was much more that could be learned, if one was prepared to look at the available fossils with an open mind. Over the next 15 years, Ediacaran organisms were reinterpreted as everything from protists, to lichens, to fungi, to colonial prokaryotes. None of this subsequent research ended up supporting Seilacher's speculations about pneu structures, or the notion that Ediacaran organisms make a new kingdom of life. But neither did this subsequent research end up totally reaffirming the previous consensus around Glaessner's work.

The lesson here is, when the scientific consensus doesn't rest on the firmest of foundations, sometimes it's enough to just shake things up. Convince researchers that there's a need for new ideas, then let the normal workings of science take their course. Easier said than done, of course. If a consensus exists at all, it's going to be difficult to shake people out of it, almost by definition. The wildness of Seilacher's speculations was helpful here, at least up to a point. His speculations were *so* far outside the consensus view that they attracted a lot of attention. But, crucially, they were also sufficiently grounded in established science that they couldn't be totally dismissed (contrast, say, the groundless wild speculations of Williamson 2009). Even skeptical attention is better than none when it comes to trying to shake up a scientific consensus.

Conclusions

There are of course other ways to attract attention to contrarian views in science and shake up the consensus, besides speculating wildly. Here are some striking examples I encourage you to look up, if you don't know them already:

- You can use strong language or other unconventional rhetoric. Think for instance of Warton and Hui (2011), which calls a then-standard data transformation "asinine" in the paper title. I wouldn't use that language myself in a paper title, but it certainly got my attention and made me read the paper closely! Further back, think of S. Gould and Lewontin (1979). That work criticizes adaptationist thinking in evolutionary biology by drawing an unusual analogy between (i) nonadaptive features of organisms and (ii) spandrels, a putatively nonfunctional architectural feature of domed medieval churches. S. Gould and Lewontin (1979) also compares adaptationist evolutionary biologists to Voltaire's Dr. Pangloss, who thinks that ours is the best of all possible worlds and that everything in it is as perfect as it could possibly be. Few scientific papers are built around architectural analogies and references to Voltaire! There are even scholarly papers on the rhetoric in S. Gould and Lewontin (1979) (Charney 1993; Queller 1995). The unconventional architectural analogy doesn't stand up to scrutiny; spandrels actually are functional (Mark 1996). But the analogy to spandrels was plausible enough to start a scientific debate.
- You can use unconventional lines of evidence. T. Clark et al. (2020) is an unusual experiment; replications are very rare in ecology. As another example, Kamath and Losos (2017) doesn't include any new data. Rather, the bulk of the paper consists of an unusually detailed review of previous work on territoriality in anoles. Few reviews in science include as many direct quotes of previous papers as Kamath and Losos (2017), or such deep dives into the nitty-gritty details of the methods of previous papers. Those unusual details were necessary in order to establish an unusual claim: that the consensus regarding territoriality in anoles rested on shaky foundations. As a third example, recall the work of Lenore Fahrig on the effects of habitat fragmentation on biodiversity (see chapter 4). Most review papers compile quantitative data on the topic itself; not many also compile quantitative data on how previous authors have written about the topic.
- You can use humor and satire. Norman Ellstrand (1983) satirized evolutionary biologists who are too quick to interpret organismal

phenotypes as adaptive, by writing an apparently serious paper hypothesizing why it's adaptive for newborns to be smaller than their parents. Here's how skillful Ellstrand's satire is: I've assigned that paper to undergraduate honors students, who didn't realize it was satire until I told them. As another example, Yoan Fourcade, Aurélien Besnard, and Jean Secondi (2018) showed that Old Master paintings predict species distributions in Europe almost as well as environmental variables do (yes, really). They could've simulated fake environmental variables to make the same point. But superimposing paintings on maps of Europe worked just as well statistically and was funnier.

- You can publish your contrarian views in some forum besides a peer-reviewed journal. Avoiding peer-reviewed journals allows you to get your views in front of an audience, without being blocked by peer review. Fretwell (1975) recalled how theoretical ecologist Robert MacArthur inspired many younger ecologists by publishing his papers in *PNAS*. MacArthur was a member of the National Academy of Sciences, and so was entitled to publish papers in *PNAS* without going through conventional peer review. MacArthur used *PNAS* to publish novel ideas that became very influential, but that would've been difficult to publish in a peer-reviewed ecology journal at the time. Another example is blogs. J. Fox (2012b) and Saunders et al. (2017) discuss blogs as a venue for scientists to share and debate a wider range of ideas, more quickly, than would be possible via peer-reviewed journals.

All these ways of attracting attention to contrarian views break the usual rules of scientific communication. Scientific papers are "supposed to be" written in a dry style, scientific papers don't ordinarily include jokes or quotations, and non-peer-reviewed communications ordinarily don't carry much weight. Breaking the rules of scientific communication arguably is justified in the rare but important cases in which the scientific consensus is dubious or incorrect. I say that even though one can certainly point to examples in which scientists successfully managed to challenge the scientific consensus without breaking any of the usual rules of scientific communication (for instance, Vellend, Baeten, et al. 2013 and Dornelas et al. 2014, discussed above).

Further, breaking the rules tends to draw negative reactions from other scientists. Some people don't like it when other people break the rules, understandably. The rules exist for a reason, and usually it's bad to break them! Robert Peters's *Critique for Ecology* (1991) used some unusually blunt language for a scientific book, and the negative reviews were unusually blunt in return (Shapiro 1993; Keddy 1992; Hengeveld 1992; Lawton 1992). The rhetoric in S. Gould and Lewontin (1979) drew a lot of criticism as well as a lot of attention (Queller 1995). *PNAS* recently cut back on opportunities for National Academy members to circumvent the usual peer-review process (P. Davis 2016), in part because some unreviewed papers later came in for substantial criticism and even ridicule (Williamson 2009; Oransky 2011). *PNAS* also cut back on circumvention because of the perception that it's unfair for some scientists but not others to be excused from normal peer review.

The bottom line is that the usual rules of scientific communication have pluses and minuses. So does breaking the usual rules—even when it's done to shake up a dubious scientific consensus. There aren't any hard-and-fast rules for breaking the rules. But I think it's important, especially for students, to realize that the rules of scientific communication can be, and have been, broken, sometimes to good effect. The rules can even be broken by students: Ambika Kamath critiqued the consensus around anole territoriality as a PhD student. In my admittedly anecdotal experience, graduate students in ecology these days generally are very keen to make sure they're doing everything "right." They're keen to stick to the usual protocol, follow a template, use best practices. They often get anxious about having to forge their own path, and especially anxious about deviating from an established path. So I think it's worth giving students a diversity of examples of successful scientific communication, including examples when rules were broken successfully.

Hopefully, if we all get a bit more comfortable with questioning scientific consensus, and a bit more familiar with the diversity of ways one can go about questioning a consensus, then dubious consensuses will get questioned more often and more effectively. When we think of questioning a scientific consensus, we often think of heated arguments—in particular, those involving senior white male faculty. S. Gould and Lewontin (1979) is perhaps the most famous example; the arguments over R. Peters (1991) provide another example. Many people find heated arguments off-putting, and for good reason. But there are plenty of other models that you could, and frankly should, follow instead. Kamath and Losos (2017), for instance, is just as critical of the consensus on anole territoriality as S. Gould and Lewontin (1979) is of adaptationism. But the authors of Kamath and Losos (2017) went out of their way to make clear that previous researchers working on anole territoriality were good scientists whose data retains substantial value, even though the authors questioned the conventional interpretation of those data. That the response to Vellend, Baeten, et al. (2013) and Dornelas et al. (2014) evolved from an argument into a collaboration is another hopeful illustration that perhaps ecologists are learning to argue less and instead disagree more effectively.

7 *Tying It All Together: The Many Roads to Generality in Ecology*

Tell Me Again: What's "Generality"?

The variety of nature presents a challenge for ecologists. Individual organisms differ from one another in ways both obvious and subtle, even if they're members of the same species living in the same location. So too do all the other things ecologists study. No two populations, species, communities, ecosystems, habitats, biomes, forest fires, heat waves, food webs, life histories, and so on are exactly alike. What, if anything, can be said *in general* about how ecological systems work? If there are generalities in ecology, do they take the form of exceptionless "laws of nature" analogous to the laws of physics? Or do they take some other form? Should ecologists even try to identify ecological generalities? If so, how? Or, if that would be a wild goose chase, what should ecologists do instead?

The variety of nature is matched by the variety of ecologists' answers to those questions. Consider the following quotes from prominent ecologists past and present:

> To do science is to search for repeated patterns, not simply to accumulate facts. —Robert MacArthur (1972, 1)

> Unlike population genetics, ecology has no known underlying regularities in its basic processes.
> —Leigh Van Valen and Frank Pitelka (1974, 925)

> The very most important thing to me, being a scientist, is to seek out unification. —John Harte (2014)

> I think of ecology as a library of well-developed case studies.
> —Tony Ives (2014)

> General ecological patterns emerge most clearly from this glorious diversity when systems are not too complicated . . . and at very large scales, when a kind of statistical order emerges from the scrum. The middle ground is a mess. —John Lawton (1999, 188)

> There are several very general law-like propositions that provide the theoretical basis for most population dynamics models. . . . Some of these foundational principles, like the law of exponential growth, are logically very similar to certain laws of physics.
> —Peter Turchin (2001, 17)

> These [previous] studies have provided more and better data on a wide range of ecological phenomena. There has not, however, been comparable conceptual progress in organizing and synthesizing existing information, producing mathematical models that are both realistic and general, and developing a body of ecological theory that can account for both the infinite variety and the universal features of organism-environment relationships. —James Brown (1997)

> Our future advances will not be concerned with universal laws, but instead with universal approaches to tackling particular problems.
> —Peter Kareiva (1997)

Clearly one generality about ecology is that all ecologists have their own opinions about generality in ecology!

Or rather, ecologists *who publicly share their opinions about generality* all have their own opinions about it. One shouldn't assume that those opinions are representative of all ecologists. Indeed, I doubt they are. My own sense is that almost all ecologists these days care about generality in some sense or other, and think or hope that it's discoverable.

But if you push ecologists on what *exactly* we mean by "generality," you get different answers. Sometimes, you also get vague answers, or differently vague answers. Now that we ecologists have mostly stopped arguing in print about what "generality" is and how to achieve it—indeed perhaps *because* we've stopped arguing—we're left with vague, pro-generality vibes.

In this chapter, I'll articulate those vibes. I identify five "roads to generality": five specific senses in which an ecological claim can be said to be "general." Different research approaches produce "generality" in different senses, useful to different researchers, depending on their goals and preferences.

I conclude by discussing the various ways ecologists have tried to discover and justify different sorts of general claims. The conclusion highlights the importance of being clear as a researcher about exactly what specific sense of generality you're trying to achieve, and why and how you're trying to achieve it. Vibes will only take you so far. If you don't know exactly what kind of generality you're aiming for as a researcher, there's a risk you won't achieve your goals.

Five Roads to Generality in Ecology

ROAD TO GENERALITY 1: META-ANALYSIS

Meta-analysis is a family of statistical techniques for quantitatively summarizing the results of different empirical studies of the same topic (Gurevitch et al. 1992; Hedges, Gurevitch, and Curtis 1999; Koricheva, Gurevitch, and Mengersen 2013). It is by far the most popular road to generality in ecology. The first meta-analysis in ecology was published in 1991; in 2022 ecologists published more than 160 meta-analyses (see chapter 1).

I'm sure many readers are familiar with meta-analysis, but in case not, here's a good example. The authors of Hoeksema et al. (2010) conducted a meta-analysis of experimental studies of the effects of mycorrhizal fungal inoculation on plant growth. The roots of many terrestrial plant species form symbiotic associations with mycorrhizal fungi. These fungi can improve plant growth because they obtain growth-limiting nutrients from the soil and trade them to the plant for photosynthate.

Alternatively, mycorrhizal fungi can act as parasites, reducing plant growth. So, what, if anything, can be said *in general* about how mycorrhizal fungi affect plant growth?

Translating this into statistical terms, we can ask: what's the average effect of mycorrhizal fungi on plant growth? The average effect is one plausible measure of the "typical" effect. We can also ask: how much variation is there around the average effect? If there's a lot of variation, we might prefer to focus on the range within which most effect sizes fall, rather than on the average effect. Finally, we can ask: what factors are associated with that variation around the average? For instance, do mycorrhizal fungi tend to improve plant growth in infertile soils, but parasitize plants in fertile soils? Together, answers to those three questions—regarding the average effect, variation around the average, and factors associated with that variation—statistically summarize what can be said, in general, about the effects of mycorrhizal fungi on plant growth.

To answer those three questions, the authors of Hoeksema et al. (2010) compiled data from 616 experiments, from more than 120 published research studies (many studies report multiple experiments). Each experiment measured total plant biomass (or in some cases, just aboveground biomass) in two treatments: an uninoculated control treatment, and a treatment inoculated with mycorrhizal fungi. One would expect the results of those 616 experiments to vary, both because of random sampling error and because the experiments differed from one another in many ways. The experiments involved different plant species, different fungal species, different soils, different settings (greenhouses, growth chambers, field sites, etc.), lasted different lengths of time, and so on. Each experiment can be summarized with a single number called an effect size. The researchers used the response ratio as the measure of effect size. The response ratio is the ratio of treatment mean to control mean, which the researchers log-transformed for purposes of analysis. A log-transformed response ratio of 0 indicates no treatment effect (i.e., equal treatment and control means), a value greater than 0 indicates that fungi increase plant growth, and a value less than 0 indicates that fungi reduce plant growth. What's the average effect size, how much variation is there around the average, and what explains that variation?

Figure 7.1 illustrates the answers to the first two of those questions, as well as how the answers changed over time as more studies were conducted. Figure 7.1, panel a, plots the untransformed response ratio from each experiment versus the rank order in which they were published (earliest is 1, second is 2, etc.). Plotting the untransformed effect sizes eases interpretation of the measurement scale. Each point gives the result of one experiment. The horizonal line highlights an untransformed response ratio of 1, indicating no effect of fungi on plant growth. The jagged lines indicate the estimated 95% prediction interval—the interval within which 95% of effect sizes are expected to fall. The width of the 95% prediction interval at any given point on the *x*-axis reflects all the studies published up to that point.

Figure 7.1, panel b, plots the estimated mean effect size (points), and the associated 95% confidence interval (jagged lines), as a function of the number of studies published up to that point. The final estimated mean effect size is approximately 1.5, meaning that on average mycorrhizal fungi increase plant biomass by 50% (treatment mean = 1.5 × the control mean). We can be quite confident that this mean truly is positive: the final 95% confidence interval ranges from 1.39 to 1.70, which is nowhere close to overlapping zero. Our confidence that the true mean is positive reflects the large number of studies that have been performed. It wasn't until more than 30 studies had been published on this topic that the lower bound of the 95% confidence interval moved comfortably far from zero. But individual effect sizes vary a lot around that nonzero mean: 25% of reported effect sizes are negative rather than positive (fig. 7.1, panel a inset), and their absolute magnitudes vary enormously. The width of the 95% prediction interval fluctuates over time as more studies are published, but it's always extremely wide. Its final width ranges from response ratios of 0.5 up to almost 9—that is, if we were to conduct a new experiment, we shouldn't be surprised if inoculation with mycorrhizal fungi cuts mean plant biomass in half, or increases it ninefold. That's a lot of variation!

Further, most of that variation is not due to random sampling error, which would decrease with increasing sample size. Rather, most of that variation represents true among-experiment variation in effect size, which meta-analysts call "heterogeneity." Different experiments on the effects of mycorrhizal fungal inoculation on plant growth produce very different results on average, and would do so even if they all had

Figure 7.1. Cumulative meta-analysis of the effect of experimental mycorrhizal fungal inoculation on mean plant biomass

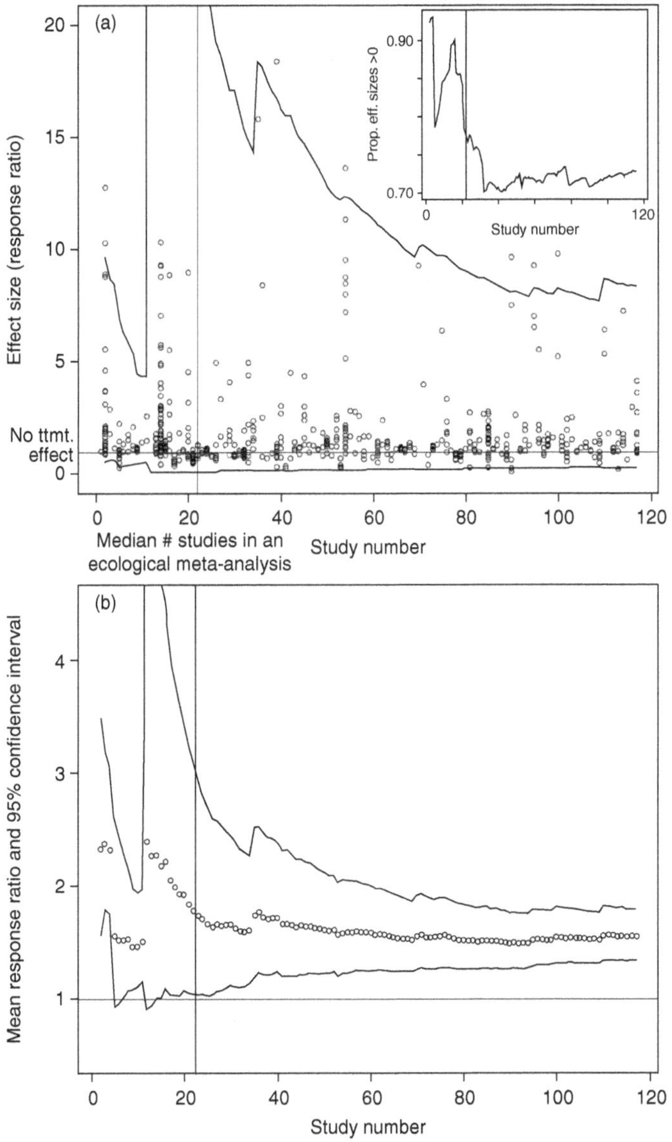

(a) Effect sizes (response ratios) as a function of study number. The first published study is number 1, the second is number 2, and so on. Each point gives one effect size. The horizontal line highlights a response ratio of 1, indicating no effect of mycorrhizal inoculation on mean plant biomass. Response ratios greater than 1 indicate a positive effect of inoculation on plant biomass. The jagged black lines are 95% prediction intervals from a cumulative hierarchical random effects meta-analysis, estimating variation in effect size within and among studies. The width of the 95% prediction interval changes from left to right along the *x*-axis, as studies are incorporated into the cumulative meta-analysis. Note that the *y*-axis is cropped to exclude a few very large positive effect sizes (the largest is >1,000), to enable clear display of the other effect sizes. The vertical line highlights the number of studies in the median ecological meta-analysis (22; Costello and Fox 2022). The inset shows how the proportion of positive effect sizes changes as more and more studies are incorporated into the cumulative meta-analysis. (b) Estimated grand mean effect size (*points*), and the 95% confidence interval for the grand mean (*jagged lines*), as a function of the number of studies incorporated into the cumulative meta-analysis. Horizontal and vertical lines as in panel a.

Source: Modified from Hoeksema et al. (2010).

Figure 7.2. Sampling error and heterogeneity in ecological meta-analyses

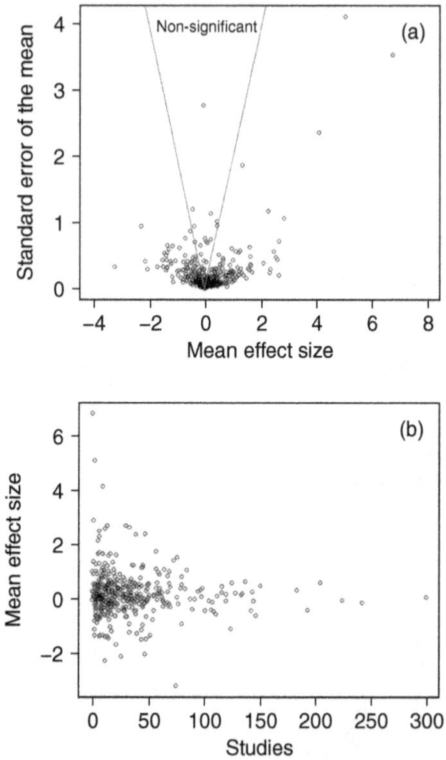

(a) Mean effect sizes and their standard errors from a compilation of more than 400 ecological meta-analyses (Costello and Fox 2022). Mean effect sizes inside the cone are within ±1.96 standard errors of zero and so do not differ significantly from zero at the 0.05 level. (b) Mean effect size versus the number of studies (primary research papers) in the meta-analysis.

massive sample sizes. What explains that heterogeneity? Probably many things, only a few of which we can identify. For instance, the authors of Hoeksema et al. (2010) found that, as expected, the average effect of fungal inoculation on plant growth is more positive in unfertilized soils than in fertilized soils. The effect of fungi on plant growth also varies with plant functional group (e.g., C3 vs. C4 grasses) and a few other ecological and methodological variables. But the many predictor variables considered by the researchers together explained just 26% of the variance in effect size.

All of which turns out to be typical. Figure 7.2 shows some results from my own reanalysis of a compilation of 467 ecological meta-analyses

Figure 7.2. (continued)

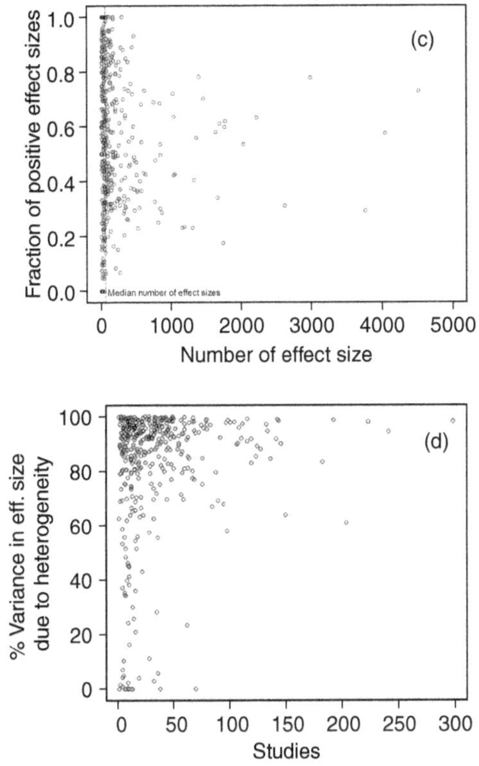

(c) The proportion of positive effect sizes in the meta-analysis versus the number of effect sizes in the meta-analysis. (d) Estimated percentage of the total variance in effect size due to heterogeneity rather than sampling error (I^2; Higgins and Thompson 2002) versus the number of studies in the meta-analysis.

(Costello and Fox 2022). Most ecological meta-analyses report a mean effect size significantly different from zero (fig. 7.2, panel a). There are now hundreds of effects ecologists have studied, about which we can now generalize with confidence regarding the sign of the average effect (although see Yang, Hillebrand, et al. 2022). That's really useful! For instance, simple theoretical models in ecology often predict the sign of some effect (see Road to Generality 2, below). Estimating mean effect sizes is one way to test whether those predictions hold on average. But other sorts of statistical generalizations are harder to come by.

For instance, what if you want to know the quantitative magnitude of the mean effect size, not just its sign? In that case, you're going to need many more studies than the typical ecological meta-analysis includes. The median ecological meta-analysis includes just 63 effect sizes from 22 studies (Costello and Fox 2022). Figure 7.2, panel b, shows that mean effect sizes from small meta-analyses vary massively. That's in part because different meta-analyses concern different effects; different effects obviously will have different means. But not *that* different, as illustrated by the fact that the points in figure 7.2, panel b, form a sideways cone shape. Mean effect sizes from large meta-analyses cluster much closer to zero than do mean effect sizes from small meta-analyses. Having only a few studies of a given topic is like having a small sample size of observations: it leads to very imprecise estimates of population parameters that could be way off in either direction (J. Fox 2022)—especially if you're sampling from a highly variable, heterogeneous population. Which ecologists invariably are when they conduct a study of pretty much any topic. And it's not just the effect of mycorrhizal fungi on plant growth that varies in sign and magnitude from one study to the next: *every* effect for which ecologists have published at least 500 effect sizes is positive in more than 20% of cases, *and* negative in more than 20% of cases (fig. 7.2, panel c). Further, almost every effect on which ecologists have published more than 50 studies is highly heterogeneous (fig. 7.2, panel d; see also Senior et al. 2016; Bebout and Fox 2024). So if there's an ecological effect that always has the same sign in every study, that effect hasn't been studied often enough.

Finally, although many ecological meta-analyses report some success at explaining variation in effect size with moderator variables, the moderator variables in ecological meta-analyses typically leave most of the heterogeneity in effect size unexplained (e.g., Davidson et al. 2017; Brisson et al. 2020; Bishop and Nakagawa 2021; Moran et al. 2021; Shanebeck et al. 2022). That's fine if the only goal is to test the sign of the effect of this or that moderator variable. But it's not fine if we want to go beyond null hypothesis testing and explain a substantial fraction of the variation we see in nature.

Ecologists working on particular research topics sometimes have documented—and lamented—the fact that research on those topics has yielded extremely heterogeneous results that have only become more heterogeneous over time as more and more studies have been conducted. For instance, consider studies of the phenotypic traits of invasive species, the factors that promote invasions, and the impacts of invasive species. These studies find effects that are weak on average, inconsistent in sign, and highly heterogeneous for reasons we can't explain (Moles et al. 2012; Jeschke et al. 2012; Hulme 2015; Guerin et al. 2018; Boltovskoy et al. 2021; Hulme et al. 2013). Further, those effects only became weaker on average, more inconsistent in sign, and more heterogeneous as more studies were conducted (Jeschke et al. 2012; Crystal-Ornelas and Lockwood 2020a, 2020b). Even after conducting thousands of studies on hundreds of invasive species, there's not much we can say with confidence in terms of statistical generalizations about invasive species or their impacts (Crystal-Ornelas and Lockwood 2020a). To the extent that invasive species ecologists think otherwise, it's probably because they're unduly focused on a few high-profile but unrepresentative cases involving just a few invasive species (Hulme et al. 2013; Guerin et al. 2018; Crystal-Ornelas and Lockwood 2020a; Boltovskoy et al. 2021). Some authors have criticized previous invasive species research for not discovering more-robust statistical generalities, and have suggested reforms that might improve matters (Hulme et al. 2013; Crystal-Ornelas and Lockwood 2020a). But here's the thing: lack of robust statistical generalizations isn't specific to research on invasive species. It's endemic to literally every single empirical research topic on which ecologists have ever conducted a meta-analysis.

I think ecologists are starting to discover the limits of meta-analysis as a road to generality. Even when we have a hundred-plus studies to go on, we're rarely able to explain much of the heterogeneity in effect size. Worse, we often don't have all that many studies to go on. Recall what I noted above, that the median ecological meta-analysis consists of just 22 primary research studies (Costello and Fox 2022). Those studies were published over a median of 20 years. That is, ecologists typically publish about one empirical paper per year on a typical ecological research topic. This isn't a recipe for rapid progress toward statistical generalizations about any of the topics ecologists study. Especially not in light of massive heterogeneity in effect size that arises from numerous, mostly unknown (or at least unmeasured) sources. You can't discover statistical generalities about any ecological topic if you don't have many studies to go on. And as examples like invasive species impacts research illustrate, you often can't discover statistical generalities even *after* you have many studies to go on.

None of which is a criticism of ecologists or meta-analysts. Ecology is a diverse, fragmented field. There are thousands of researchers working on hundreds of topics. There's no way to drastically reduce the number or diversity of topics ecologists study so as to force all ecologists to focus on the same few topics. And I doubt we would want to reduce the number of topics on which ecologists work, even if we could. Nevertheless, a field in which a diverse range of ecologists work on a diverse range of topics is not one that's going to make rapid progress toward statistical generalizations about any one of those topics. At least, the field won't make much progress on any of those topics within the professional lifetime of a typical ecologist. If our only source of generality in ecology is meta-analysis, we're not going to end up saying much in general about much of anything.

So if we're going to achieve generality in ecology about anything besides the signs of the mean effect sizes or the signs of the effects of a few moderator variables, we can't rely solely on meta-analyses. We'll never have enough studies of all the topics ecologists study. And even if we did, the results of all those studies would be highly heterogeneous, and we wouldn't be able to explain most of that heterogeneity. Meta-analyses are well worth doing. But it's also worth complementing them with other, less popular roads to generality. "Generality" has other meanings that aren't captured by meta-analytical summary statistics.

ROAD TO GENERALITY 2: BUILD A SIMPLE MODEL OF ONE OR TWO KEY PROCESSES

Robert MacArthur pioneered this road to generality. At the core of this approach is what C. S. Holling (1966) called a "strategic" model and Richard Levins (1966, 422) called a "precise general" model: a mathematical model that describes one or two key processes in a simple way, and that assumes away other processes entirely. (For other discussions of this road to generality, both pro and con, see Odenbaugh 2005; Ishida 2007; Evans et al. 2013.)

For instance, MacArthur and Wilson's (1963, 1967) theory of island biogeography models changes in species richness on islands as a function of rates of two processes: immigration from the mainland of species not already present on the island, and extinction of species already present on the island. Both of these rates were assumed to be simple functions of island species richness (fig. 7.3). Other examples of this road include (but aren't limited to) MacArthur's limiting similarity theory of interspecific competition (MacArthur and Levins 1967), simple predator-prey models like the Rosenzweig–MacArthur model (Rosenzweig and MacArthur 1963), the marginal value theorem of optimal foraging (Charnov 1976), simple resource competition models (Tilman 1982), simple models of optimal life history (D. Cohen 1966), and the May–Wigner theorem relating stability to complexity in simple ecosystem models (May 1972).

This approach provides generality in the sense that the model applies approximately to various different systems, though not exactly to any system. This is what ecological theoreticians mean when they refer to such models as "capturing the essence" of the process being modeled. Many different study systems might share the same "essence," even if they differ from one another in other respects.

For instance, the determinants of species richness on any particular real island will not be exactly as described in MacArthur and Wilson (1963, 1967); see J. Brown and Lomolino (2000) for one review among

Figure 7.3. The theory of island biogeography

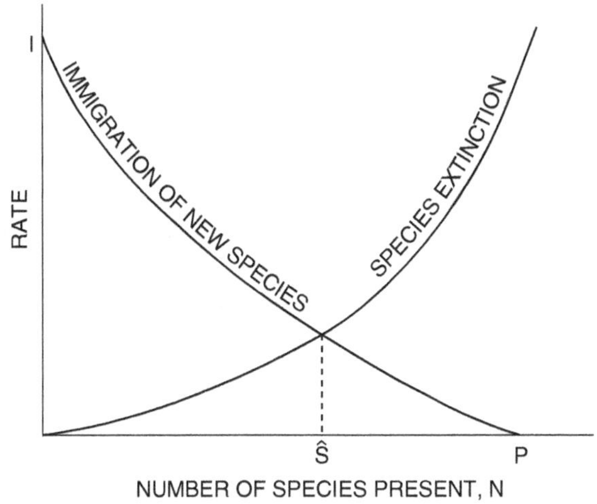

Source: Modified from MacArthur and Wilson (1967).

many. Speciation on the island might contribute new species, contrary to MacArthur and Wilson's assumption of no speciation. Species might vary in their probabilities of colonizing and going extinct, so that per species rates of colonization and extinction depend not just on the number of species currently on the island, but also on which particular species are currently there. Rates of colonization and extinction might vary over time for all sorts of reasons, rather than remaining constant. And so on. But the model might nevertheless be a good qualitative approximation; its qualitative predictions often hold, even though its assumptions never do. For instance, island biogeography theory predicts that equilibrium island species richness will increase with increasing island area and colonization rate—qualitative predictions of the signs of the associations between pairs of variables.

Of course, the quality of the approximation provided by the model is an empirical question. The MacArthurian approach works best when the process or processes included in the model are among the most important determinants of whatever is being modeled, meaning that the effects of the included processes are not swamped, or too much modified, by effects of other processes. I suspect that's why the MacArthurian approach has had many of its greatest successes in areas in like optimal life-history theory and optimal foraging theory (see Stephens and Krebs 1987 and Stearns 2000 for review, including discussion of the sense[s] in which optimality theory can be considered "successful"). Perhaps simple models of optimal life-history strategies and optimal foraging often work well empirically because evolution by natural selection is a potent force. Life-history strategies and foraging behaviors that deviate substantially from the fitness-maximizing ones are likely to be selected out.

Fortunately, the modeler has some power to ensure that a simple model includes the few most important processes. For instance, the theory of island biogeography defines the processes of interest—colonization and extinction—at a very "high" level. In situ speciation aside, it really is the case that the only way the species richness of an island can change is via colonization of species not already present, and extinction of species already present. Of course, many "lower-level" factors besides island size and distance from the mainland might affect the precise rates at which colonization and extinction occur (e.g., whether the species concerned can fly or swim, whether they compete for food, how often the island is hit by hurricanes). But so long as the model correctly describes how colonization and extinction rates depend on island size and distance from the mainland (at least approximately), it implicitly summarizes the net effects of all the underlying unmodeled factors. By implicitly summarizing the effects of many underlying "low-level" processes, MacArthur and Wilson minimized the number of processes omitted from their model entirely.

You also can try to force the MacArthurian approach to work by making it conditional, and therefore somewhat less general. That is, build a simple model of the one or two processes that will be most important, assuming that specified background conditions hold. Then develop different simple models appropriate to different background conditions. For instance, Murdoch et al. (2002) shows that population cycles in nature fall into a small number of different classes, defined by the length of the cycle period relative to the generation time of the species. Which class a given cycle falls into depends on whether the species concerned has a specialized diet or not. This suggests the need for (at least) two different simple models to explain population cycles: one for dietary specialists, another for generalists. The problem is, results like those in Murdoch et al. (2002) are rare in ecology. Ecologists mostly haven't been very good at identifying the relevant background conditions, perhaps because there are usually too many relevant background conditions for the conditionalization approach to work in ecology. Useful "periodic tables," analogous to the periodic table of chemical elements, are rare in ecology (see Pianka et al. 2017 for one attempt). The bits of nature that ecologists study just can't often be classified into groups with large among-group variation and little within-group variation. And if you condition on too many things, you're back to a bunch of unique special cases.

When the one or two key processes are not overwhelmingly important compared to those the model ignores, the MacArthurian approach can still yield general conceptual insight, and is vital as a check on one's pre-theoretical intuitions (Servedio et al. 2014). But the model predictions often will fail to hold. Also, they may be difficult to test in an informative way. If your model makes only one or a few qualitative predictions (e.g., variable *y* will increase with variable *x*), it can be difficult to distinguish a model that is approximately right in the essentials, despite being incorrect in detail, from a model that's wrong in the essentials. And if a model is supposed to apply only approximately to any particular empirical system, often it is hard to test by checking the correctness of its assumptions (instead of, or in addition to, the correctness of its predictions), because those assumptions aren't supposed to hold exactly. For this reason, it's often best not to test the qualitative predictions of simple MacArthurian theoretical models. Instead, it's better to test quantitative predictions of more complicated, system-specific versions of those simple models. The simple model identifies and defines the general phenomenon that you're testing for, and shows how different system-specific cases can all be viewed as instances of that general phenomenon. The more complicated model is more testable because it explicitly describes the biology of the specific system in which it's being tested. For instance, Jennifer Gremer and Larry Venable (2014) developed and tested a more detailed, system-specific version of the simple optimal seed germination model in D. Cohen (1966).

ROAD TO GENERALITY 3: LOOK FOR CONSISTENCY ACROSS DIFFERENT MODELS

Levins (1966) is the classic statement of this road to generality. While discussing strategies for model building in population ecology, Richard Levins remarks:

> There is always room for doubt as to whether a result depends on the essentials of a model or on the details of the simplifying assumptions. . . . Therefore, we attempt to treat the same problem with several alternative models each with different simplifications but with a common biological assumption. Then, if these models, despite their different assumptions, lead to similar results we have what we can call a robust theorem which is relatively free of the details of the model. Hence our truth is the intersection of independent lies. (423)

I think of Levins's idea of robust theorems as a modification of the MacArthurian approach of simple models, discussed above. Rather than relying on one simple model to capture the shared "essence" of many different real-world cases, Levins suggests relying on the shared features of several different models to capture that "essence."

A good example of a "robust theorem" in ecology comes from modeling how species' geographic ranges will shift in response to climate change. Imagine that you know where a given species is found now and want to predict where it will be found in the year 2100. How would you predict that? Well, there are various approaches. You could build a statistical model that relates the species' probability of presence or absence from a given location to measurements of relevant environmental variables at each location. Then you could predict where the species will be found in 2100, by using a climate model to predict where the environments in which it's currently found will be located in 2100.

But this modeling approach could go wrong for various reasons. Perhaps the species' current geographic range isn't determined primarily by current environmental conditions, and so the species' range in 2100 shouldn't be expected to be determined by environmental conditions in the year 2100. Perhaps interactions with competitors or predators are important determinants of species' geographic range limits and climate change will alter the outcomes of those interactions (A. Davis et al. 1998). Perhaps barriers to dispersal will prevent the species' geographic range from shifting so as to track changing environmental conditions. And so on. All of this motivates an alternative modeling approach: to build and parameterize a mechanistic model rather than a statistical model. That mechanistic model will mathematically describe how species' rates of survival, reproduction, and dispersal depend on environmental conditions, the presence or absence of competitors and predators, and other relevant factors. Hopefully, that model will account for how dispersal, changing species interactions, and other factors will combine with climate change to alter species' geographic ranges in the year 2100.

It's not clear yet which modeling approach is best. But one thing we can do is compare the predictions of different approaches. Predictions that are shared across multiple approaches are robust predictions in Levins's sense. The authors of Morin and Thuiller (2009), Kearney, Wintle, and Porter (2010), and Fordham et al. (2018) all studied the robustness of predicted range shifts to different modeling approaches.

As another example, I have used Levins's notion of robust theorems in my own work on the spatial synchrony of population cycles. Spatially separated populations of cycling species often cycle in lockstep: all populations increase, and decrease, at the same time (Liebhold, Koenig, and Bjørnstad 2004). That's likely because population cycles are easily synchronized by dispersal—movement of organisms from one population to another (Vasseur and Fox 2009). My collaborator David Vasseur and I developed three different simple models of how population cycles might be synchronized. All three incorporated the same key ingredients: dispersal, population cycles, and random fluctuations in the external environment. We included environmental fluctuations because spatially synchronized environmental fluctuations also can generate spatial synchrony. The three models differed in the details of how they described these ingredients (Vasseur and Fox 2009). All three models made the same qualitative prediction: dispersal synchronizes population cycles, random environmental fluctuations don't. That prediction held in an experiment as well. Because the same behavior occurred in four different cases that shared the same key ingredients but differed in other respects, we argued that any natural system sharing these key ingredients should be expected to exhibit the same behavior as well.

But perhaps that argument was unjustified! There is a large philosophical literature developing, revising, and critiquing Levins's notion of a "robust theorem." Questions of interest include (i) what exactly "robustness" is, (ii) whether robust theorems really are more likely to be true, and (iii) whether robust theorems are desirable or useful for other reasons, besides their likelihood of being true (Orzack and Sober 1993; Weisberg 2006; Wimsatt 2007; Odenbaugh 2011; Parker 2011; Knuuttila and Loettgers 2011; Odenbaugh and Alexandrova 2011; Justus 2012).

ROAD TO GENERALITY 3A: ASK WHETHER DIFFERENT MODELS PRODUCE THE SAME PROBABILITY DISTRIBUTION OF OUTCOMES

A special case of the previous road to generality is when the "robust theorem"—the shared essential core of the overlapping models—is statistical in nature (Frank 2014). This special case crops up often in ecology, so it's worth singling out for discussion.

For instance, consider a normal distribution, the so-called bell curve. Normal distributions are quite common. Figure 7.4 shows just a few haphazardly chosen examples, from ecology as well as other fields. Many,

Figure 7.4. Examples of normal distributions

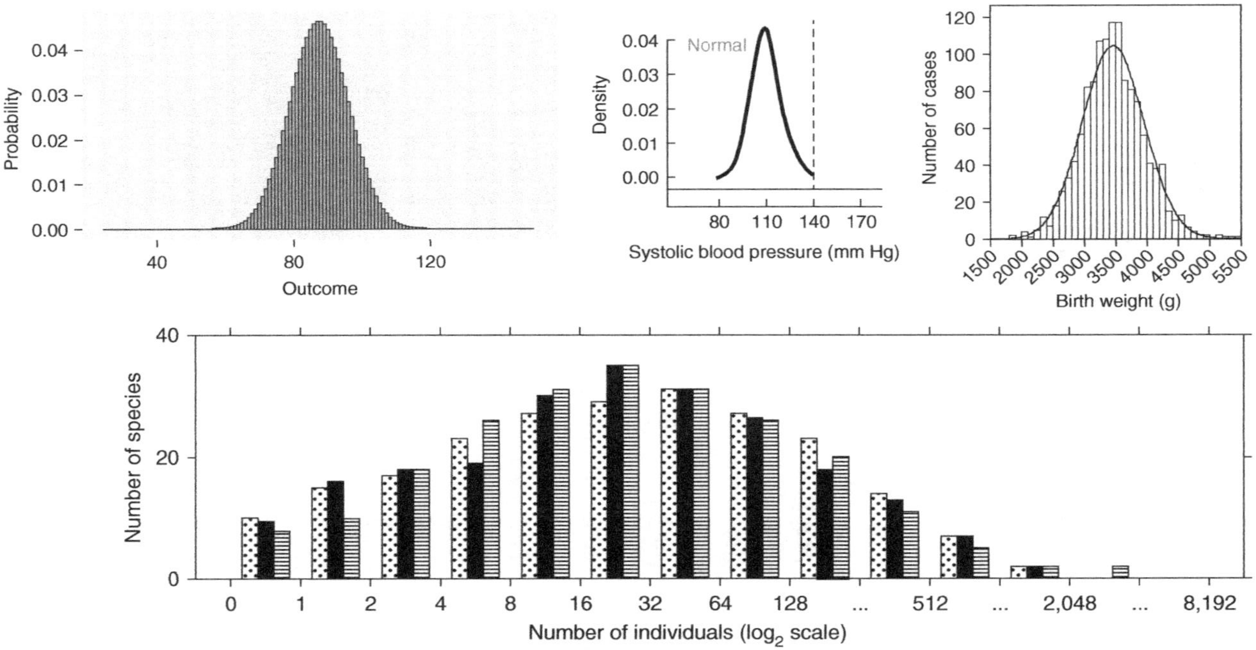

Top row, left to right: The distribution of the sum of 25 six-sided dice, the distribution of systolic blood pressure for nonhypertensive US women ages 18–39 (J. D. Wright, Hughes, et al. 2011), and the distribution of birth weights of 1,295 infants born at a hospital in Berlin, Germany (Pfab et al. 2006). *Bottom*: The observed distribution of log-transformed abundances of tree species in a 50-hectare plot on Barro Colorado Island, Panama (*black bars*), and two statistical distributions (*stippled and striped bars*) with parameters chosen so as to match the observed distribution as closely as possible (McGill 2003).

many variables exhibit approximately normal distributions. Why is that? What do they all have in common?

What they all have in common is that their observed values reflect the approximately additive effects of many different contributing factors. The central limit theorem tells us that if you sum up a bunch of random numbers, then sum up a new bunch of random numbers, and so on, the resulting sums will be normally distributed. This is true even if the random numbers are sampled from different distributions, even if the random numbers are somewhat correlated with one another, and no matter the order in which you add them up. That's why the sums of the rolls of a bunch of dice are normally distributed (fig. 7.4, top row, left panel). But it's also why human birth weights are normally distributed (fig. 7.4, top row, right panel). Human birth weight is affected by many different factors that vary from one baby to the next—the baby's genotype at each of many different genetic loci, various aspects of the mother's health, and so on. There are complex genetic, physiological, developmental, and biochemical causal pathways by which all those factors shape human birth weight. But the net *effect* of each factor is to add or subtract a bit of mass to or from the baby's birth weight. Same story for human blood pressure (fig. 7.4, top row, middle panel): it depends on many different factors, each of which has the effect of raising or lowering blood pressure by a bit (in some cases, more than a bit). Add up the effects of all those factors in each of many different people, and you end up with a normal distribution of blood pressure.

And it's the same story for species abundances, so long as you log-transform them (fig. 7.4, bottom panel). Population growth is a multiplicative process, not an additive one: you multiply current abundance by per capita growth rate over some time period to figure out what the abundance will be at the end of that time period. Which means that population growth is additive on a log-transformed scale. Many different factors (various competitors, various predators, various parasites, various weather variables, etc.) will increase or decrease per capita growth rate a bit, or sometimes by more than a bit. Add up the effects of all those factors on a log-transformed scale, and you end up with an approximately normal distribution of log-transformed species abundances (called a "lognormal" distribution). You end up with an approximately lognormal distribution—even if the precise factors at work differ from one species or one community to the next, have different effects from one species or one community to the next, and so on. Presumably this is why observed species-abundance distributions typically look roughly like lognormal distributions, and why any plausible mathematical model of species-abundance distributions (and many implausible ones) produces a roughly lognormal distribution (Wilson et al. 2003; Wilson and Lundberg 2004; McGill, Etienne, et al. 2007; Hammal et al. 2015; Baldridge et al. 2016; Frank and Bascompte 2019).

Normal distributions are common because they are hard to avoid (Frank 2014); the normal distribution is a "statistical attractor." For instance, any plausible model of population growth will be multiplicative, so the modeler would have to make quite specific (and unrealistic) assumptions for it *not* to predict an approximately lognormal distribution of species abundances. Identifying a commonly observed pattern in empirical data as a statistical attractor thus can explain why it's commonly observed. Generality in the sense of "would be observed in many different cases, regardless of case-specific details" leads to generality in the sense of "often observed."

ROAD TO GENERALITY 4: UNIFY A BUNCH OF DIFFERENT SPECIAL CASES UNDER A SHARED UMBRELLA

This approach provides generality in the sense of unification. Different special cases are all subsumed under the umbrella of a more general case.

The theory of evolution by natural selection is the most important example of this road to generality. For instance, consider changes over time in (i) beak sizes of Darwin's finches and (ii) antibiotic resistance in bacteria, and the causal factors driving those changes. At first glance, those two cases apparently have nothing to do with one another. Darwin's finches with different beak sizes differ in their abilities to consume different sizes of seeds (Grant and Grant 2014). This is for biomechanical reasons, having nothing to do with the biochemical reasons why some bacteria are resistant to antibiotics. But the evolution of beak size in Darwin's finches and the evolution of antibiotic resistance in bacteria are in fact related: they are two classic examples of evolution by natural selection. In both cases, there is heritable variation in phenotypic traits affecting fitness.

An ecological example of a unifying theoretical framework is modern coexistence theory (Chesson 2000; Ellner et al. 2019). Under a few (somewhat restrictive) assumptions, modern coexistence theory identifies all possible classes of mechanism by which competing species can coexist in the long run without any one of them driving the others to extinction. Modern coexistence theory also provides mathematically precise definitions of those classes. Modern coexistence theory, like most unifying theoretical frameworks, thus is descriptive rather than predictive: it tells us which real-world systems fall within the same space of possibilities, and it carves up the space of possibilities in a useful way. But the theoretical framework doesn't itself make any predictions about where real-world systems will fall within that space.

Unifying theoretical frameworks aid the search for generality in at least two ways. First, they make possible useful empirical comparisons, allowing us to ask and answer questions we wouldn't otherwise be able to ask. For instance, to be able to statistically compare the strength of natural selection on different phenotypic traits using meta-analysis (Kingsolver et al. 2001), you first have to have a unifying abstract concept called "natural selection," defined sufficiently precisely to be measurable or estimable in a wide range of cases. Unifying theoretical frameworks thus complement case studies. Second, unifying theoretical frameworks ensure that no possibilities go unrecognized. If ecologists did not have modern coexistence theory, they likely would not be able to exhaustively enumerate all possible classes of coexistence mechanism. On the *Dynamic Ecology* blog, I once argued that research on "metacommunities" (networks of spatially separated local communities connected by dispersal; Leibold et al. 2004) had been held back due to failure to enumerate all possible classes of metacommunities (J. Fox 2013c). I further argued that mistakes could have been avoided if metacommunity ecology had a unifying theoretical framework (J. Fox 2013c).

The challenge with this road to generality is that the unifying umbrella might cover so many different special cases as to be vacuous. Imagine trying to come up with a unified general theory of "stuff"—one that applies to all the stuff in the world and explains why stuff is the way it is. That's silly, because the term "stuff" is too broad and vague. It covers too many unrelated things that have nothing to do with one another. So any general unifying theory of stuff would have to be vacuous ("Stuff happens."). No ecologist has ever proposed such an *obviously* overbroad and vague concept as "stuff." But there are examples of ecological concepts that are at least *arguably* overbroad and vague, as we'll discuss in chapter 8.

ROAD TO GENERALITY 5: IDENTIFY FRUITFUL ANALOGIES BETWEEN APPARENTLY UNRELATED CASES

Ecological theoretician Tony Ives (2014) articulated this road to generality well in an interview:

> As a theoretical ecologist, you might think that I'm motivated by general laws. But I don't find general laws very interesting. I really like solving problems. If I'm using theory and not looking for universal patterns or universal laws, then people ask what the hell am I doing? My answer is that I think of ecology as a library of well-developed case studies. If you've come across something in your own system, you can go to the bookshelf, pull out a book—a case study—and read it. And maybe there's some insight that inspires you to look at your system differently. This makes case studies very useful for ecology. . . . I wouldn't argue that all case studies are singular and have nothing to do with each other. Clearly there are a lot of themes that cut across different systems.

This approach provides generality in the form of fruitful analogies between apparently unrelated cases. These cases might all come from within ecology. But to my mind, the most valuable analogies are those between ecology and other scholarly fields. Fruitful analogies between different fields both enable rapid progress within a field and allow existing theory from one field to be reinterpreted and applied to another field.

Consider the example of spatially synchronized population cycles, discussed above (Vasseur and Fox 2009). Spatially separated, cycling populations linked together by movement of organisms are just one among many examples of "coupled oscillators" in various domains of science (Strogatz 2004). Other examples include synchronous pulses of light from lasers, the pacemaker cells of your heart (the synchrony of which keeps your heart beating rhythmically), and fireflies that all flash in time with one another. Similar or even identical mathematics can be used to describe the causes of synchrony in all these cases, although the interpretation of the mathematical parameters is case-specific. Other examples of this road to generality in ecology and allied fields include the idea of "critical transitions" in both natural systems and human societies (Scheffer 2009), reinterpretation of population genetic theory as a theory of community ecology (Hubbell 2001; Vellend 2016), the application of reaction-diffusion equations from chemistry to spatial ecology (Cantrell and Cosner 2004), and reinterpretation of the Price equation from evolutionary biology so that it can be applied to analogous problems in ecology and other fields (J. Fox 2006; B. Clark 2010).

There are two main challenges to following this road to generality in ecology. First, loose analogies are a dime a dozen and not very helpful. In ecology, think for instance of "broken stick" models of the species-abundance distribution (MacArthur 1957). These models predict the shape of species-abundance distributions like the one in the bottom panel of figure 7.4 by analogy with breaking a stick. The tropical tree community on Barro Colorado Island has divided up the resources available to the trees (space, light, etc.), with the abundant species having claimed more of the resources than the rare species. Analogously, we might think of the length of a stick as representing the total amount of resources available to a community. Breaking the stick into a bunch of pieces, according to some algorithm, is analogous to dividing those resources between a bunch of species, the abundance of each of which is proportional to the length of its piece. It turns out that this isn't a helpful analogy. Various broken-stick models, based on various breakage algorithms, have been proposed. For instance, a breakage algorithm that involves breaking the stick into pieces sequentially (rather than simultaneously) is loosely analogous to an ecological process in which species colonize a habitat sequentially rather than all colonizing at once (Tokeshi 1993). Broken-stick models provide poor fits to most observed species-abundance distributions (Alroy 2015). Worse, the analogy between stick breaking and community ecology is just too loose to be useful. If a given breakage algorithm fits the observed data well (or poorly), all you've learned is that . . . your stick-breaking algorithm fits the observed data well (or poorly). The analogy between your stick-breaking algorithm and the processes that determine species abundances is just too loose for us to learn anything about *why* it does, or doesn't, fit the data. It's rather like asking an AI image generator for an image of a sunset. Whether or not the image resembles a real sunset, it doesn't teach you anything about why sunsets look the way they do.

The second challenge to following this road to generality is that, in my admittedly anecdotal experience, many ecologists just don't care much about this sort of generality (J. Fox 2019a). Most of us want generalizations about ecology, especially the statistical generalizations provided by meta-analyses. Relatively few of us care about generalities linking ecology to other fields. One common criticism from peer reviewers of my work on spatial synchrony of population cycles is that spatially synchronized population cycles are statistically rare. Only a minority of species exhibit population cycles (Kendall, Prendergast, and Bjørnstad 1998), and not all cycling species exhibit spatial synchrony. That makes spatially synchronized population cycles uninteresting or unimportant in the eyes of these reviewers, who care more about generality in the sense of what is statistically common than about generality in other senses. That spatially synchronized population cycles are an ecological analog to coupled oscillators from physics, chemistry, physiology, and animal behavior is a mere curiosity in the eyes of these reviewers. To my mind, though, the analogy facilitates rapid progress in ecological studies of spatial synchrony. For instance, you can learn a lot about spatially synchronized seed production in trees by reinterpreting models of coupled atomic spins from physics (A. Noble et al. 2018). The close analogy between coupled oscillators in ecology and other fields also is just profound to me. The same math describes the synchronization of masting in trees and the synchronization of magnetic dipole moments of atomic spins (Strogatz 2004; A. Noble et al. 2018)! That is just amazing and satisfying to me, for reasons having nothing to do with how common masting is relative to unsynchronized reproductive dynamics. Many interesting and even profound things are rare in a statistical sense. This is one reason why it's useful to recognize different roads to generality. Nobody is obliged to share my deep satisfaction at the analogies between spatial synchrony of population cycles and other coupled oscillators. But they at least ought to be able to appreciate where that deep satisfaction comes from.

How to Combine Different Roads to Generality (or Not)

Different specific senses of generality, and the research approaches by which they are achieved, are complements, not substitutes. They often are more useful in combination than on their own. Discussion of this point could have appeared in chapter 3, on complementarity of research approaches. But I'm going to keep it here for the sake of continuity.

Roads to generality 1 (statistical summaries) and 2 (simple theoretical models) can be a powerful combination, when the statistical summary identifies a clear-cut pattern such as a strong correlation between two variables. A clear-cut pattern that seems to hold in many different cases suggests the need for a theoretical explanation that is independent of case-specific details (MacArthur 1972; J. Brown 1997). An example from evolutionary ecology: metabolic rate and many other organismal attributes scale with body size, in a manner often well described by a statistical relationship known as a power law (G. West, Brown, and Enquist 1997; J. Brown 1997). The same power law relationship fits data from species as different as lizards and whales. This general statistical pattern motivated the development of a theoretical explanation that ignores many species-specific biological details to focus on a key mechanism that applies to many different species: evolutionary optimization of branched circulatory systems (G. West, Brown, and Enquist 1997; J. Brown 1997).

Roads 2 and 4 can be a powerful combination: unify or summarize the effects of various "lower-level" processes or factors under a "higher-level" umbrella (road 4), and then build a simple model of those high-level processes or factors (road 2). Classical population genetic models such as the Wright–Fisher model provide a textbook example of combining roads 2 and 4. Classical population genetic models are specified in terms of a small number of high-level parameters such as selection coefficients, population sizes, mutation rates, and migration rates that summarize the effects of lots of complicated, system-specific biology.

Combining road 5 (fruitful analogies) with road 1 (statistical summaries) and/or 4 (unifying theoretical frameworks) can be a powerful combination. But it's rarely attempted in ecology, and the power depends on the analogy being sufficiently precise to support the development of a unifying theoretical framework. The idea is to identify an analogy between disparate cases, then formalize the analogy sufficiently to unify those cases under a shared theoretical framework, and/or allow comparative statistical analysis.

The work of Mark Vellend (2016), mentioned earlier in this chapter as an example of road to generality 5 (fruitful analogies), may be better thought of as combining roads 4 and 5. Drawing a sufficiently precise analogy between community ecology and evolutionary biology allows the unifying theoretical framework of evolutionary biology to be reinterpreted as a unifying theoretical framework for community ecology.

An example of combining roads 5 and 1 is work treating food webs as topological "networks": abstract sets of "nodes" (here, species) connected by "links" (here, between predator species and the prey species they eat). Other networks in this abstract sense include the World Wide Web (web pages are the nodes, hyperlinks are the links), electrical grids (power stations and substations are the nodes, transmission lines are the links), protein interaction networks (the proteins produced by an organism are the nodes; they are linked if they interact biochemically), and scientific citation networks (scientific papers are the nodes, citations are the links). The authors of Dunne, Williams, and Martinez (2002) and Milo, Shen-Orr, et al. (2002) compared the topological properties of food webs (e.g., frequency distribution of links per node; measures of whether nodes fall into highly interlinked "clusters") to those of other networks. In some respects, food webs are typical of other networks, suggesting that these topological properties of food webs may not require an ecology-specific explanation (indeed, perhaps suggesting that some of those topological properties are statistical attractors that many networks exhibit because it's hard to avoid exhibiting them). In other respects, food webs differ from other networks, suggesting that the processes determining food-web structure are importantly different from the processes structuring other networks. In particular, some networks are optimized by human engineers, or natural selection, to perform some function. Other networks, including food webs, are not. Networks that are optimized to perform a function are likely to have structural features reflecting that function (Milo, Shen-Orr, et al. 2002; Milo, Itzkovitz, et al. 2004). There's no particular reason to expect non-optimized networks or networks optimized for a different function to share those structural features.

Note that not all roads to generality can be combined usefully. For instance, roads to generality 2 and 3a (simple models and statistical attractors) play poorly together. A statistical attractor retains no information about the details of the underlying contributing factors that generated it. Ecologists often have forgotten this—for instance, attempting to distinguish between alternative simple ecological models of population growth and species interactions by comparing their ability to reproduce observed species-abundance distributions. This is an ineffective approach because pretty much every plausible simple model of population growth and species interactions generates an approximately lognormal species-abundance distribution, and so is about equally consistent with the observed data (McGill, Etienne, et al. 2007; Wilson and Lundberg 2004; Hammal et al. 2015).

Quit Telling Other Ecologists That There Are No Roads to Generality, or That They're Taking the Wrong One

Ecologists don't need to pick a single road to generality. This is a case of different tools for different jobs, not a case where it's either necessary or desirable to select the best option from a diverse set of candidates. Recognizing the diverse senses of "generality" in ecology resolves multiple long-standing debates.

AN ANSWER TO SHRADER-FRECHETTE AND MCCOY (1993)

In an influential book, Kristin Shrader-Frechette and Earl McCoy (1993) argued that ecology (particularly applied ecology) should, and does, follow a "logic of case studies." Shrader-Frechette and McCoy were concerned with how ecologists can make justified inferences about specific cases (e.g., how to conserve a specific endangered species) when general principles are vague or inapplicable. The authors conceded that their proposed logic of case studies does not support scientific generalizations, but they set this concern aside as irrelevant to their goal of making case-specific inferences. This represents a missed opportunity on their part. There are several ways case studies provide a basis for various different senses of "generality." My taxonomy also reveals how different roads to generality can contribute to case-specific inferences. For instance, one can use meta-analysis (road to generality 1) to improve the accuracy of effect size estimates from the studies included in the meta-analysis ("shrinkage estimation," Efron and Morris 1977; J. Fox 2022).

AN ANSWER TO LAWTON (1999)

In a famous paper, John Lawton (1999) claimed that evolutionary biology, population ecology, and macroecology have generalities, but community ecology does not and so should no longer be pursued. My taxonomy of roads to generality provides a way to push back against this claim. Lawton recognized that both evolutionary biology and population ecology have historically made heavy use of road to generality 4 (unifying theoretical frameworks), while macroecology—the study of patterns in the distribution and abundance of species at large spatial scales—has made heavy use of road to generality 3a (statistical attractors). Lawton is correct that community ecology historically has made little use of either of those two roads to generality. But he implicitly assumes that (i) those are the only two roads to generality, and (ii) neither of those roads could ever be open to community ecology. Both of those implicit assumptions are incorrect.

Regarding assumption (i), statistical summaries such as meta-analyses (road 1) are just as possible in community ecology as in any other field. Linquist et al. (2016) shows that commonly used covariates in ecological meta-analyses (variables associated with taxonomy, habitat type, and spatiotemporal scale) explain just as much variation in community-level variables as they do in population- or ecosystem-level variables. Community ecology also has simple theoretical models and robust theories (roads 2 and 3). For instance, predators reduce the abundance of the herbivores they prey on, which in turn increases the abundance of the plants the herbivores consume. Such "trophic cascades" can be explained qualitatively with simple food-chain models (Oksanen et al. 1981). Trophic cascades also are ubiquitous in experimental studies (Shurin, Borer, et al. 2002), likely because they are robust—they occur in many different ecological scenarios (e.g., Abrams 1993).

Regarding assumption (ii), recall that Mark Vellend (2016) developed a unifying theoretical framework for those parts of community ecology concerned with diversity and coexistence of competing species. Interestingly, that framework is precisely analogous to the unifying framework of evolutionary biology: it is based on four fundamental "forces," each analogous to one of the four fundamental forces of microevolution (selection, drift, mutation, migration). Community ecology, far from being inherently "messier" than evolutionary biology, can be viewed as conceptually analogous to evolutionary biology if ecologists choose to view it that way (Vellend 2016). Community ecologists can of course choose not to do so, but at the cost of giving up the right to complain about the messiness of their own field. They are no more entitled to complain about the inherent messiness of their field than someone who ties his own shoelaces together is entitled to complain that he can only walk slowly.

Conclusions

My response to Lawton (1999) generalizes beyond community ecology. The nature and extent of any generalities we discover in ecology are partially under our own control. Generality, or the lack thereof, is not a property just of nature itself. Rather, it is a joint property of nature and the way ecologists choose to study it. Depending on how nature is, generality in some specific senses might not exist or might be difficult to discover. But that's just another way of saying that, depending on what questions we choose to ask and how we choose to pursue them, generality in some specific senses might not exist or might be difficult to discover. Recognizing that the generalities we discover are partially under our control increases the number of ways the search for generality might succeed. If one road to generality turns out to be a dead end in a particular case, or turns out not to lead anywhere worth going, ecologists have both the opportunity and the responsibility to choose another road.

8 *The Downsides of Diversity*

Diversity is not always a good thing.

Ecologists know this well, or at least we should. As discussed in chapter 5, BEF experiments sometimes report negative selection effects (e.g., Jiang 2007; D'Andrea et al., n.d.). These effects arise when initially diverse communities come to be dominated by highly competitive species that don't contribute much to the desired function. "Highly competitive" and "high functioning" are two different things that don't always go hand in hand (Jiang, Pu, and Nemergut 2008; S. Wang et al. 2021). Negative complementarity effects are possible as well (D'Andrea et al., n.d.).

There are various other mechanisms by which negative effects of diversity can arise, but they don't all have analogs in BEF experiments. Diversity of goals and approaches isn't an unmitigated good for the field of ecology. But we can't learn much about the reasons why not by analogy with BEF experiments. In this chapter, I'll discuss how the diversity of goals and approaches can hold back progress in ecology. At the end, I'll discuss what, if anything, can or should be done about such cases.

Diversity of Terminology: The Ecological Tower of Babel

What's the best way to measure floompstruts?

You can't answer that question, because you don't know what a floompstrut is. No one does, including me. Because I just made it up. "Floompstrut" is a silly way to make a simple point: you can't put any numbers on a thing unless you know what the thing is.

Ecologists never try to measure floompstruts. But they do try to measure (or count, or otherwise quantify) many other things that aren't all that much better defined than floompstruts are. Now, every word has a least a bit of vagueness or ambiguity in its definition (Wittgenstein 2010). A bit of vagueness or ambiguity doesn't usually cause serious problems in everyday conversation. But many important ecological terms are *so* vaguely defined that it's not at all clear how to operationalize them—how to define them in a sufficiently precise and practical way that empirical researchers can put numbers on them. The result is that the term gets operationalized in a diverse range of ways by different ecologists. Who then get different numbers, which we struggle to compare, interpret, and reconcile.

It's an ecological version of the Tower of Babel. Everyone uses the same terms to mean different things, so we all talk past one another while getting frustrated and confused.

Or worse, maybe nobody actually knows what a particular term means. Maybe everyone just uses the same vaguely defined term without even realizing it's vaguely defined.

Don't believe me? Here's what's surely a *very* incomplete list of ecological terms with disputed or vague meanings. (It's surely a very incomplete list because it includes only the disputed or vague terms I could think of off the top of my head right this second.)

Biodiversity. I'm old enough to remember when this term was just starting to come into use in ecology. I recall attending a biodiversity workshop as a grad student. We spent the entire workshop trying and failing to agree on a definition. The term "biodiversity" was coined around 1986 by Walter G. Rosen, a staff member at the National Research Council, the advice-giving branch of the National Academy of Sciences. Rosen helped organize the 1986 National Forum on BioDiversity. The forum was an "exercise in consciousness-raising," in Rosen's phrase (Takacs 1996, 38). That exercise needed a new, pithy term that could encompass everything that everyone valued about living nature. The new term had to be vague and all-encompassing, so as to avoid hard choices about conservation priorities, trade-offs, and triage. Which didn't mean that definitional issues could be dodged entirely. David

Takacs (1996) asked various prominent ecologists to define "biodiversity" and got various definitions that shared little beyond being broad and vague. The interviewed ecologists defined "biodiversity" as everything from "the number of species and the uniqueness of species" (48) to "diversity at all levels of organization" (48) to "importance: which areas do we have to concentrate on, or which groups are more 'important' in terms of preserving than others" (48).

Alpha, beta, and gamma diversity. Robert Whittaker (1960) introduced the terms "alpha," "beta," and "gamma" diversity into ecology. "Alpha diversity" refers to diversity within sites, "beta diversity" refers to diversity among sites (because species richness, composition, and abundance may vary among sites), and "gamma diversity" refers to the total diversity of a set of sites. Ecologists have spent the last 65 years debating exactly how to define and measure alpha, beta, and gamma diversity. For instance, should these terms be defined so that gamma diversity is the sum of alpha and beta diversity, or the product of alpha and beta diversity? Different definitions of "alpha" and "beta diversity" are related to one another, yet can produce quite different values. Various reviewers have tried to sort out this mess by classifying different measures of alpha and beta diversity, and recommending the best measures of each for various purposes. Trouble is, different reviewers have produced different classifications and recommendations (Lande 1996; Jost 2006; Jurasinski, Retzer, and Beierkuhnlein 2009; Tuomisto 2010; Baselga and Leprieur 2015; Ricotta and Feoli 2024; Ricotta 2017).

Niche. "Niche" is perhaps the most important term in community ecology besides "community" itself. But ecologists don't agree on how to define it, and never have (Sales, Hayward, and Loyola 2021). Rather, researchers working on different topics in ecology all use the word to mean different things. For instance, "ecological niche modeling" doesn't really have much to do with the use of the term "niche" in coexistence theory (Chesson 2000; Thuiller 2024).

Stability. I once wrote a blog post reviewing twenty different definitions of "stability," and I omitted a bunch (J. Fox 2012a). Grimm and Wissel (1997) compares the proliferation of definitions of "stability" to the Tower of Babel.

Ecosystem health. Scientists, governments, and nongovernmental organizations have been trying to restore ecosystems to "health" (or "integrity") for decades, even though nobody knows how to define or measure ecosystem health. Ecosystem health, like beauty, is in the eye of the beholder, apparently. O'Brien et al. (2016, 722) reviews definitions and indicators of ecosystem health for aquatic ecosystems, finding that "few studies clearly defined ecosystem health and justified the choice of indicators." O'Brien and coauthors more or less threw up their hands, suggesting that the only practical approach was for different investigators "to define and measure health on a case by case basis" (722). Without wanting to deny that the definition of "health" is contested even when applied to humans, imagine if medical doctors all felt free to define and measure the health of their own patients in their own way, on a case-by-case basis. I doubt that this would inspire much confidence in their patients!

Habitat fragmentation. Lindenmayer and Fischer (2007) criticizes this term as a "panchreston"—a term used to cover such a broad and vaguely defined range of phenomena as to be meaningless. Echoing and sharpening this point, Fahrig (2017a, 2017b) argues that habitat fragmentation research had produced numerous misleading, incorrectly interpreted results by using a single term indiscriminately to refer to habitat loss as well as to "fragmentation *per se*" (breaking up contiguous habitat into noncontiguous patches of the same total area as the original contiguous habitat).

Importance and intensity of competition. Welden and Slauson (1986) attempts to distinguish the "importance" of competition from its "intensity." The authors defined "importance" as "a physiological concept, related directly to the well-being of individual organisms but only indirectly and conditionally to their fitness," distinct from "intensity," which the authors defined as "primarily an ecological and evolutionary concept, related directly to the ecology and fitness of individuals but only indirectly to their ecological states" (23). The paper was very influential, despite—or perhaps because—of the fact that subsequent researchers operationalized "importance" and "intensity" in different, contradictory ways (Freckleton, Watkinson, and Rees 2009). Freckleton, Watkinson, and Rees (2009) notes that "the importance of competition is 42"—a reference to the central joke of

The Hitchhiker's Guide to the Galaxy, in which the meaning of "life, the universe, and everything" is revealed to be "42."

How much of a problem is it for ecologists to use the same terms to mean different things, or for a term to be so vague that nobody knows what it means? I think it depends.

I think it's a problem when researchers try to measure or operationalize a vaguely defined concept. It's a waste of time for everybody to come up with their own operationalization and then argue about which is best. Different operationalizations will give different answers, and there's no way to decide which answer is correct, or best, because nobody *really* knows what they're even trying to measure. Arguments about which operationalization is best are a distraction. In fact, the only thing all sides agree on—that the argument is worth having—is exactly what they ought to be questioning. That's why I agree with Freckleton, Watkinson, and Rees (2009) that it was a waste of time for researchers to propose and implement many different, contradictory indices of the "importance" and "intensity" of competition. That wasted effort could've been avoided if researchers hadn't tried to operationalize such vague concepts in the first place.

It's also a problem when researchers lump together multiple distinct concepts under one definitional umbrella without realizing that they're doing so. That's why I agree with Fahrig (2017a, 2017b) that the literature settled on an overly negative consensus regarding the effects of habitat fragmentation on biodiversity. In practice, "habitat fragmentation" was defined so as to lump together effects of habitat fragmentation per se with effects of habitat loss. It would've been more informative to separate these effects. To be clear, there are often good reasons to lump concepts together under some broader definitional umbrella. But you always want to do the lumping knowingly, to make sure it makes sense to lump them together. And you can use your knowledge of the narrower concepts that you've lumped together to aid interpretation of the umbrella concept.

In other cases, I don't think use of the same term to mean different things has created much of a problem. For instance, I think "stability" researchers are mostly pretty good about making clear precisely what they mean by "stability" in any particular case. Using "stability" to mean various things doesn't seem to lead to major confusion or wasted research effort. Same for "niche"—I don't think anyone is confused, or that any mistaken scientific conclusions are drawn, because ecologists doing, say, ecological niche modeling and coexistence theory use the term "niche" in different ways. Ecological niche modelers all know what they mean by "niche," and people doing coexistence theory all know what *they* mean by "niche," and everybody knows that the two groups of researchers use the word in different ways.

Others have argued that there are cases in which vague definitions actually are *helpful* to scientific progress. Waters (2013) discusses how research progress in genetics has if anything been accelerated by the fact that the term "gene" is only vaguely defined, so that the term has different referents for different authors. The biological reality of genetics is too complicated and variegated to be fully captured by any one precise definition of "gene." Agreeing on a single precise definition would retard research progress by forcing researchers to ignore or misdescribe some aspects of genetics. Speciation research in evolutionary biology provides another example. Evolutionary biologists have learned a lot about how speciation works (Coyne and Orr 2004). I suspect that's in part because of, rather than despite, the fact that the term "species" has various proposed definitions (biological species concept, phylogenetic species concept, etc.), all of which have some real-world cases to which they apply only fuzzily, or not at all (Coyne 1992; Coyne and Orr 2004). Sometimes it's helpful for our definitions to be a little bit vague and allow for a few edge cases. Hodges (2008) discusses additional ecological examples of vague terms that aid research progress.

In cases where it is a problem for ecologists to use the same term to mean different things, or to use a vaguely defined term, what can be done about it? I'm not sure. I do know what *can't* be done: attempts to propose standardized terminology are doomed to fail. I say this because many ecologists have made such attempts (e.g., Fauth et al. 1996; Colautti and MacIsaac 2004; Stroud et al. 2015; Soto et al. 2024, Fusco et al. 2024). Those proposals were all well intended. But the next such proposal that succeeds will be the first. No one who proposes standardized terminology has any power to force others to even learn about their proposal, much less adopt it. And so the proposals end up merely growing the pile of conflicting definitions, rather than shrinking it (Hodges

2008). As illustrated by the fact that proposals to standardize terminology in this or that area of ecology tend to be followed a couple of decades later by . . . proposals to standardize terminology in those same areas of ecology (e.g., Fauth et al. 1996; Colautti and MacIsaac 2004; Stroud et al. 2015; Soto et al. 2024, Fusco et al. 2024).

Researcher Degrees of Freedom

"Researcher degrees of freedom" refers to the fact that researchers have to make many decisions when analyzing a dataset. We've already discussed the freedom of choice you often have when deciding how to operationalize ecological concepts. But your choices don't end there; far from it. Even after you've collected the data, you have choices to make. Should you transform the response variable, and if so, how? Should you discard that outlier? Should you use a Bayesian approach, and if so, what prior should you use? Which covariates should you include in your statistical model? Should you fit a parametric nonlinear curve to your data, or a nonparametric smoother, such as a cubic spline? And so on. Many of those decisions aren't obvious, cut-and-dried choices. In any given case, there are various more or less reasonable choices one might make.

So, as a data analyst, you have some freedom of choice. That's a form of diversity: given the same dataset, different analysts will make different choices. How much do those choices matter? You might hope that any reasonable set of analytical choices will lead to similar bottom-line conclusions. But is that hope justified?

One way to find out is to give the same dataset to each of many analysts. Ask them all to answer the same scientific question by analyzing that dataset however they think best. Then compare their answers. One of the first papers to do this was Silberzahn et al. (2018). The authors gave each of 61 analyst teams a large dataset on red cards in European professional soccer matches. The dataset included information on many possibly relevant predictor variables. The authors asked the analyst teams whether soccer referees are more likely to give red cards to darker-skinned players. No two teams came up with exactly the same answer. How much the answers varied depended on which aspects of the answer you considered. Almost every team estimated the same sign of effect (darker-skinned players are more likely to receive red cards).

Figure 8.1. Results of two "many analysts/one dataset" studies

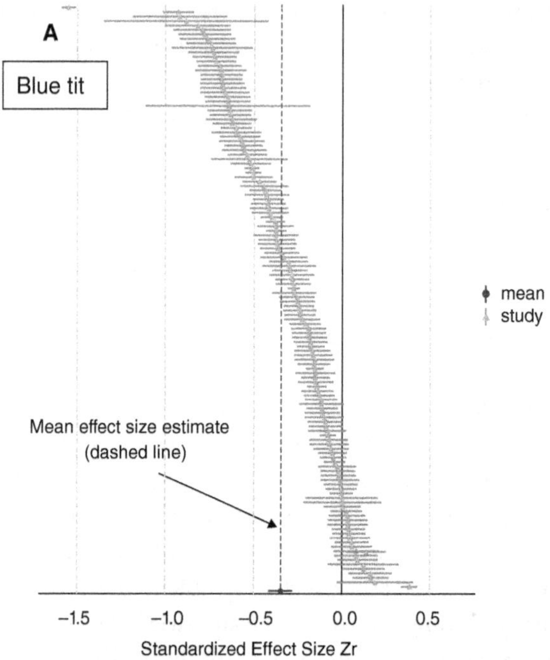

Each point and associated horizontal line are the estimated mean effect size (*z* score) and associated 95% confidence interval from one analyst team. Analyses in (A) concern an experiment on clutch size and nestling growth in blue tits. Analyses in (B) concern observational data on grass cover and *Eucalyptus* forest regeneration. In each panel, the vertical dashed line marks the grand mean effect size, averaging over all analyses. The vertical solid line marks a mean effect size of zero.

Source: Modified from E. Gould et al. (2025).

There was somewhat less agreement on statistical significance: 69% of teams agreed that the effect of skin color was statistically significant. But the teams varied hugely in their estimates of the quantitative magnitude of the effect.

Subsequently, other investigators have published "many analysts/one dataset" studies in other fields. Some of which have obtained broadly similar results to Silberzahn et al. (2018). In particular, E. Gould et al. (2025) reports two "many analysts/one dataset" studies in ecology (fig. 8.1). The authors found broad among-analysts agreement as to the sign of the effect in question (negative for one dataset, near zero for the

Figure 8.1. (continued)

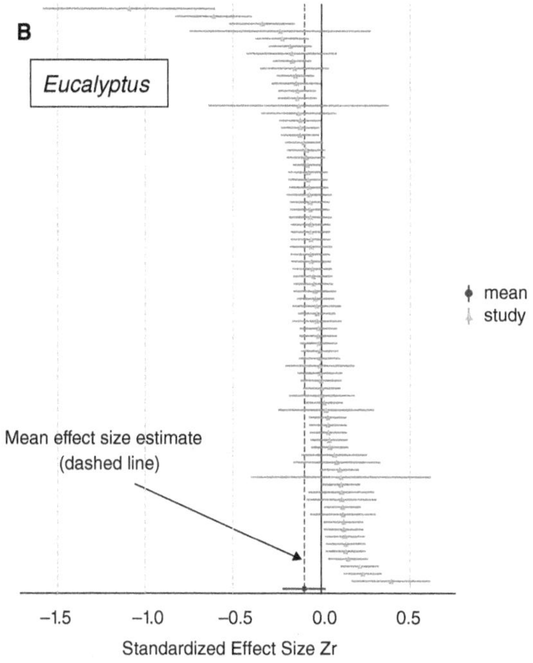

other), but moderate or substantial disagreement among analysts as to quantitative magnitude. However, other "many analysts/one dataset" studies find absolutely massive among-analysts variability not just in the magnitude of the effect, but in its statistical significance and *even in its sign* (Huntington-Klein et al. 2021; Breznau et al. 2022). Worse, nobody who's looked has been able to find *any* predictors of which analysts will get which answers. It's not that, for example, the analyses judged to be best by peer reviewers all tend to be similar, whereas the bad analyses are all bad in their own ways (Silberzahn et al. 2018; E. Gould et al. 2025). Indeed, reviewers tend to disagree a lot as to which analyses are best, or as to the quality of any given analysis (E. Gould et al. 2025). It's not that analysts with more statistical expertise get different answers from those with less (Silberzahn et al. 2018). It's not that analysts have different preexisting beliefs about what the answer will be, and then all tend to obtain whatever answer confirms their beliefs (Silberzahn et al. 2018).

How worried should you be about diverse analysts obtaining diverse answers to the same scientific question from the same dataset? I can see arguments for being very worried, and arguments for being fairly blasé. One reason for being blasé is that ecologists typically care much more about the sign of an effect than about its quantitative magnitude. Most (not all) analysts usually agree about effect sign (Silberzahn et al. 2018; E. Gould et al. 2025; but see Huntington-Klein et al. 2021; Breznau et al. 2022). One reason for being worried is that ecologists arguably *ought* to start caring about effect magnitudes (Popovic et al. 2024). But ecologists can't start caring about effect magnitudes if we all estimate wildly different effect magnitudes from the same data.

But however worried you are, I think you have to agree that it's not *good* that diverse analysts often obtain diverse answers to the same scientific question from the same data. You can't select among these diverse answers in order to find the best one if reviewers disagree as to which answers are best. Nor are the diverse answers complementary to one another—that is, we don't learn anything from the entire set of answers together that we couldn't learn from considering each answer on its own. For instance, it's not clear that the average of the analysts' diverse answers is any better than any analyst's individual answer. Nor do different answers do different jobs. They're all answers to the same scientific question, using the same data.

Conclusions

The take-home lesson from this chapter is straightforward: diversity of ideas and approaches is only a problem when researchers don't realize the ideas and approaches are diverse. It's a problem when we don't realize that we're all talking past one another because we're using the same term in different ways. It's a problem when we don't realize that we've lumped together different concepts under one umbrella. It's a problem when we don't realize that different statistical models would give us different estimates. In order to effectively harness diversity, we have to know it's there.

9 *It's Not Just Ecology*

I said up front that ecology suffers from two related anxieties: we don't know what it is, and we don't know how to do it. In response, I argued that those anxieties reflect the diversity of ecologists' goals and approaches. This diversity should be a source of strength, not a weakness, if only ecologists recognize and harness it.

A person who feels anxious sometimes will find it reassuring to learn that others feel the same way. It helps to know that you're not alone. So, in the hopes that what goes for individuals also goes for an entire scientific field, next I'll show that ecology isn't alone. Every other scholarly field with which I have some passing familiarity suffers from similar anxieties. There's a diversity of views as to what each field's goals should be and how those goals should be pursued, but that diversity isn't being harnessed. And people who know those fields better than me suggest the same solutions I've suggested here: embrace a diversity of ideas and approaches, particularly via complementarity, but also by selecting for new approaches that improve on old ones.

Throughout this chapter, I rely on the views of scholars in other fields, since I lack firsthand knowledge.

Pure Mathematics

I'll use the field of pure mathematics as my first example, and the one I'll discuss in greatest detail. Pure mathematics is a *very* different field than ecology in many respects. So it might surprise you to learn that it suffers from the same two anxieties as ecology: disagreement among its own practitioners as to what it is and how to do it.

In his famous essay, "The Two Cultures of Mathematics," Fields Medalist William Timothy Gowers (2000) identifies two kinds of mathematicians: problem solvers and theory builders (see also Dyson 2015). Problem solvers and theory builders disagree with one another *as to the very point of mathematics*:

> The "two cultures" I wish to discuss will be familiar to all professional mathematicians.
>
> Loosely speaking, I mean the distinction between mathematicians who regard their central aim as being to solve problems, and those who are more concerned with building and understanding theories. . . . If you are unsure to which class you belong, then consider the following two statements:
>
> (i) The point of solving problems is to understand mathematics better.
>
> (ii) The point of understanding mathematics is to become better able to solve problems.
>
> Most mathematicians would say that there is truth in both (i) and (ii). . . . However, many, and perhaps most, mathematicians will not agree equally strongly with the two statements. (Gowers 2000, 1).

Gowers goes on to argue that mathematics as a whole needs both problem solvers and theory builders, because some subfields of mathematics demand problem solving while other subfields demand theory building.

Trouble is, not everyone agrees that mathematics needs both kinds of mathematician. At the time Gowers was writing, theory building was widely viewed as the "core" activity of mathematics. At least, it was viewed that way by the theory builders themselves. By their own lights, theory builders are the ones doing the important work of identifying general principles and unifying different branches of mathematics. They see "mere" problem solvers as reveling in problem-specific—and therefore unimportant—details.

Gowers goes on to note that, if it were in fact the case that problem solvers really did just solve unrelated problems one by one, their work really would be unworthy of the attention of other mathematicians. After all, so much mathematical research is published every year that no one can pay any attention to more than a tiny fraction of it. In order to be worthy of any attention at all, a mathematical problem has to be generalizable in *some* way, has to have *some* sort of connection to other mathematical problems. Fortunately, as Gowers notes, his own subfield—combinatorics—isn't just a bunch of unrelated problems. Rather, there are general principles that link different problems, but they're *different kinds* of general principles than those that theory builders seek. Theory builders make the mistake of assuming that the "road to generality" that they travel themselves is the *only* road to generality. They mistakenly think any mathematician who doesn't travel their preferred road to generality must not care about generality at all.

Gowers concludes by noting that there are negative consequences for the field of pure mathematics when theory builders and problem solvers misunderstand and devalue one another's goals and approaches. Mathematicians from one culture will have difficulty fairly evaluating the work of mathematicians from the other culture (for instance, when making hiring and tenure decisions). Prospective research students who are best suited to one culture might find themselves pressured to work in the other, and so fail to achieve their full potential. Finally, mutual misunderstanding between the two cultures constitutes a missed opportunity for complementarity. Some mathematical problems can't be solved without collaboration between members of both cultures.

Political Science

Let's turn from pure mathematics to another field that might seem very different from ecology, and from pure mathematics, but actually isn't: political science.

Like ecology, political science has often struggled to harness the diversity of its practitioners' approaches and goals. But don't take my word for it; take the word of political scientists Scott Ashworth, Christopher Berry, and Ethan Bueno de Mesquita (2021, 1):

> The rise of formal theory and the credibility revolution are two of the great developments in social science over the past half century. With these advances, the potential for productive dialogue between theory and empirics has never been greater.

> So it is distressing that, in political science, theory and empirics appear to be drifting apart. Ironically, these two developments, which should be drawing scholars together, have instead been dividing them.

Let's unpack this. By "formal theory," the authors mean causal models that address "large-scale" questions relevant to many different polities, such as the causes of civil wars. Theoreticians in political science often turn to comparative observational evidence to test their formal theories—for instance, regarding the frequency and intensity of civil wars in different countries. Those countries vary along multiple dimensions predicted to cause civil war by different causal models. By "credibility revolution," the authors mean the revolution in statistical methods for credibly inferring causality from observational data. Various such methods have been developed over the past few decades, all of which have stringent data requirements (see chapter 4). These stringent data requirements constrain the sorts of scientific questions that political scientists can answer with credible causal inference methods.

Which leads to conflict. On the one side are political scientists who would rather ask big important questions relevant to all polities, even though those questions are impossible to answer definitively because the available observational data don't allow reliable causal inference. On the other side are political scientists who would rather take full advantage of the power of new statistical methods, and who don't see much point in asking big questions if those big questions can't be answered definitively. As Ashworth, Berry, and Mesquita (2021, 1) put it:

> On one side are scholars concerned that the pursuit of credible causal estimates is displacing the canonical goal of understanding important political phenomena. The papers we write, they argue, seem no longer to be about the questions that motivate us. "Why does this important thing happen?" has been replaced by "What is the effect of x on y?" . . . They see adherents of the credibility revolution as dismissive of what can be, and indeed what has been, learned by empirical scholars employing other approaches. The credibility revolution, they hold, unnecessarily limits the scope of evidence that is considered legitimate. . . .

> On the other side are scholars who have embraced the credibility revolution, arguing that much of the canonical quantitative work in political science offered only what Gerber, Green, and Kaplan (2014) call "the illusion of learning." For these scholars, there is no point in tackling questions that cannot be answered well. We should instead focus on questions accessible to credible research designs.

Ironically, the shoe is now on the other foot for formal theorists. Today, they value asking big questions, even if those questions can't be answered with much rigor. But back in the day, they were the ones accused of overvaluing rigor. Here are Ashworth, Berry, and Mesquita (2021, 2) on the earlier division between verbal theorists and advocates of formal mathematical theory:

> In the role of today's critics of the credibility revolution were those worried that a fetishization of mathematical elegance was distracting political scientists from the goal of generating insights that were genuinely useful for explanation or suitable for empirical assessment. Green and Shapiro (1996, p. 54) lamented that "empirical progress has been retarded by what may be termed method-driven, as opposed to problem-driven, research." What is interesting or useful, critics asked, about narrow models built on assumptions that bear, at best, only a distant relationship to reality? . . .
>
> Lined up to oppose such critics were those arguing that formalization allows scholars to avoid errors of logic and achieve greater transparency. Responding to Walt, Powell (1999, p. 98) argued, "Even if tightening the connections between assumptions and conclusions were all that formal theory had to offer, this would be a very important contribution." Cameron and Morton (2002) point to three virtues of formalization: seeing with clarity which assumptions drive which results, avoiding mistakes of logic through rigor, and achieving a kind of unity or coherence by eschewing hypotheses that depend on contradictory assumptions.

As Ashworth, Berry, and Mesquita (2021, 3) note, these divisions retard progress because political science as a whole is failing to harness the complementary strengths of researchers with different goals and approaches:

The moment is ripe to draw these two groups back together. Formal theory and the credibility revolution are natural partners that, together, can support a richer and more productive dialogue between theory and empirics than has ever before been possible in political science.

However, as a discipline, we are not currently prepared to realize this potential. Empiricists and theorists alike are too quick to dismiss one another's enterprise. We all need a better framework for thinking about how the two fit together. Each side needs to better understand what kind of knowledge the other is trying to create, and how they go about it. Only with this understanding will theorists see how to make their models genuinely useful to empirical inquiry and empiricists see how to structure their research in ways that speak to theoretically meaningful questions.

Geography

The examples of pure mathematics and political science concern fields that harbor long-standing disagreements regarding research goals and approaches. But those disagreements haven't proven so serious as to threaten the existence of those fields. Their practitioners do the sorts of things that make a scholarly field coherent and viable. They attend the same conferences, they're members of the same scholarly organizations, they publish in the same journals, they're members of the same college and university departments, they teach in the same undergraduate programs, they supervise graduate students in the same graduate programs, and so on.

The same used to be true of the field of geography. But it isn't any longer—at least, not in the US. Geography emerged as a scholarly field roughly 150 years ago. In the 19th and early 20th centuries, most major US universities had departments of geography. That began to change in 1948, when Harvard University closed its Geography Department. Harvard President James Conant's stated reason for the closure was that geography was "not a university subject." The true reasons are numerous and disputed (N. Smith 1987; Mountz and Williams 2023). But whatever happened at Harvard, it can't explain the subsequent closures of large numbers of geography departments in both the US and the UK from the 1960s onward (Haigh and Freeman 1982; Sacks 2015; Cox 2021).

Geography declined as a discipline, likely because it never defined itself as a discipline in the first place. Historically, much scholarship in geography consisted of "regional" geography: diverse, location-specific case studies. Those case studies asked different questions, used different methods, and overall had little to do with one another. After World War II, such case studies became less valued by universities and funding agencies, but geography had nothing to replace them with (Hurst 1985). Geography was a diverse interdisciplinary field in terms of its questions and methods. Indeed, it was *so* diverse that it struggled to articulate reasons why it shouldn't be broken up, and different bits of it merged with different disciplines (e.g., geology, sociology; Hurst 1985; Dorling 2019; Cox 2021).

The fate of geography provides a cautionary tale: it's possible for a scholarly field to be so intellectually diverse that it ceases to be a field at all. I'm tempted to argue that ecology in the US was fortunate to avoid the same fate as geography, thanks to the influence of Robert MacArthur, E. O. Wilson, Richard Levins, and others, along with the rise of the environmental movement in the 1960s (Miller and Reed 1965).

Anthropology

Anthropology provides a second cautionary tale of the worst that can happen to a field that fails to fully harness the intellectual diversity of its practitioners. Geography struggled to articulate a shared vision of itself. Anthropology has the opposite problem: two opposing visions.

Anthropology was founded in part in opposition to spurious biological ideas that purportedly justified racism (W. Davis 2021). That opposition was enormously successful not just within anthropology but in society at large. Thomas Gossett (1997) even went so far as to suggest that pioneering anthropologist Franz Boas might have done more than anyone in history to fight racial prejudice.

But one unfortunate side effect of anthropology's opposition to racism seems to be permanent reflexive suspicion of any attempt to incorporate biology into our understanding of human culture (Engelke 2018). Anthropology is divided into two subfields: cultural anthropology and physical or biological anthropology. And as best I can tell as an outsider, "divided" really is the right word. The division runs sufficiently deep that it doesn't merely hold the field back, but threatens its continued existence as a single field of scholarship. A division so deep as to lead to the breakup of a leading anthropology department into separate cultural and physical departments is a *very* deep division (Gibbs 1998; Shea 1998; Leslie 2000). There are widespread, long-standing concerns among anthropologists regarding this division; those concerns are associated with repeated calls to heal it (see Johnston and Low 1984; Fuentes 2016; Zeder 2018 and references therein). A few of those calls have even been acted upon. Several years after the Stanford University Anthropology Department split into separate cultural and physical departments, the university administration forced the departments to re-merge. The re-merger was forced on the grounds that the two departments would teach better, and pursue better scholarship, if they worked together to harness their intellectual diversity (Jaschik 2007).

Fundamental Physics

Fundamental physics has two different theoretical frameworks describing different physical forces: general relativity and quantum mechanics. The trouble is that those frameworks are mutually incompatible. They can't both be right, so at least one of them needs to be modified or replaced, if fundamental physics is to have a unified "theory of everything" free of contradictions. But the search for a unified "theory of everything" has stalled. Many physicists argue that the search has stalled because of a lack of diversity of ideas. Over the past few decades, the field has channeled most of its collective research effort into pursuing just a couple of possible ways of unifying general relativity and quantum mechanics (primarily, string theory). Those ways haven't panned out, leaving the field at a loss for ideas as to how to move forward (Woit 2006; Smolin 2007; Carroll 2006; Hossenfelder 2018; Butterfield 2019). This example illustrates the risk of "selecting" too strongly in favor of a single research approach, and the value of intellectual diversity as insurance against that risk.

Evolutionary Biology

In my view, evolutionary biology is a model example of a scholarly field taking full advantage of the diversity of its practitioners' interests, skills, methods, and goals. The Modern Synthesis unified Charles

Darwin's pioneering verbal ideas, the then-new field of genetics, mathematical theory, observational data from fields as different as ornithology, biogeography, and paleontology, and laboratory experiments (Huxley 2010). This unification involved bringing together lines of research that weren't merely different, but in some cases opposed to one another. For instance, in the years immediately after the rediscovery of Mendel's genetic results, Mendelian genetics was widely seen as an alternative to Darwin's ideas rather than a complement to them (Berry and Browne 2022).

Ever since the Modern Synthesis, there's been ongoing disagreement as to whether it needs to be revised, expanded, or even replaced, so as to incorporate ideas and data that it ignores or that contradict it. Modern Synthesis cofounders Ernst Mayr and J. B. S. Haldane disagreed as to whether its assumption of "beanbag genetics" makes it unable to accommodate the more complicated realities of inheritance and development. Stephen Jay Gould (1980) declared the Modern Synthesis "effectively dead" because it overemphasizes natural selection and cannot accommodate the roles of chance events and constraints in shaping macroevolutionary outcomes. Lynn Margulis argued that the Modern Synthesis should be replaced with "symbiogenesis"—the hypothesis that evolutionary novelties arise primarily via cooperative genetic mergers (Margulis 1991; O'Malley 2015). The authors of Laland et al. (2015) called for an "extended evolutionary synthesis" that would incorporate developmental biology and "niche construction" (organisms' abilities to alter their own environments, thereby altering the selection pressures they experience). Several authors have suggested that DNA methylation and the CRISPR-Cas9 system falsify key tenets of the Modern Synthesis (e.g., D. Noble 2013).

My own view is that evolutionary biology has been remarkably successful at taking in an ongoing influx of new ideas and approaches, using them in ways that complement existing ideas and approaches, and selecting against those that aren't fruitful or don't stand up to scrutiny. For instance, symbiogenesis is now universally accepted as explaining the origins of mitochondria and chloroplasts, but rejected as an explanation for the origins of other organelles. When you look at how thoroughly and effectively evolutionary biology adopted new gene-sequencing technologies, it's hard to see a field that's either hidebound or too quick to chase the latest passing fad.

Conclusions

As these examples hopefully make clear, ecology's twin anxieties—that we don't know what it is, and that we don't know how to do it—do not originate in ecology's infamous messiness. Yes, ecology is a fuzzily defined scholarly field. And yes, the things it studies are hugely variable in their properties. But even though the numbers and symbols studied by pure mathematicians don't vary in their properties, pure mathematicians have the same disagreements as ecologists do over their field's goals and research approaches. And even though the particles studied by physicists also don't vary in their properties, progress in fundamental physics has ground to a halt because the entire field bet big on a single research approach and lost that bet. Conversely, anthropologists, political scientists, and evolutionary biologists all study things that are just as variable in their properties as the things we study. But yet, anthropologists and political scientists worry constantly that their fields are split into opposing camps that ought to cooperate, while evolutionary biology adapts in response to ongoing inputs of new ideas and approaches.

The lesson here is that it's up to us. The field of ecology is whatever we ecologists want it to be.

10 *The Hedgehog and the Fox*

We saw in the previous chapter that every scholarly field faces the same challenges ecology does in harnessing a diversity of research goals, ideas, and approaches. So what can we ecologists do to meet those challenges? In particular, what can we do that we aren't already doing, or aren't doing very much of?

That's a good question. I know I'm supposed to have a good answer. Because that's the only satisfying way for a science book to end: by telling the reader how to solve the scientific problems the book has identified. Andrew Hendry (2014) calls this the "baby–werewolf–silver bullet" rule. When writing a scientific paper (or in my case, a book), you first identify the goal you want to achieve (the baby). Then you identify the threat to achieving that goal (the werewolf). Then you explain how to overcome the threat (the silver bullet). Following the baby–werewolf–silver bullet rule gives your paper or book a satisfying narrative structure, like a good novel or film. I've identified the baby (harnessing a diversity of ideas, goals, and approaches), and I've pointed out the werewolf (more than one, actually). So now I should give you the silver bullet—explaining what ecologists should do differently in the future, in order to fully harness their diverse ideas, goals, and approaches. Otherwise, I've written the scientific equivalent of *No Country for Old Men*, the Joel and Ethan Coen film that (in)famously ends not with a climax or resolution, but with seemingly random events unrelated to those that came before.

I'll give it my best shot. But I'll warn you up front: I don't know that you're going to find my silver bullets all that satisfying. In my own defense, I think there's a good reason for that, which I'll discuss at the very end.

New Goal: Forecasting

In chapter 1, I noted that ecologists' research goals haven't really changed during my professional lifetime. Descriptive research has remained by far the most common sort, explanatory research has remained comparatively rare, and so on. But ecologists' research goals might be about to change in a big way, and I think that's exciting.

Historically, ecologists have hardly ever done forecasting—at least, not outside of a few specialized subfields such as fisheries ecology. Which is probably just as well. If you're still working on describing natural variation, and you can't explain much of the variation you describe, you're not really in a position to produce forecasts anyone would want.

However, that's been changing slowly but steadily throughout my career. In 1993, only about 1 out of every 500 papers in ecology journals concerned forecasting enough to mention the term (or related terms such as "forecast," "forecaster," etc.) in their title or abstract (fig. 10.1). Thirty years later, it's up to 1 out of 50, and the growth shows no signs of slowing down (fig. 10.1). Now, not all ecology papers with words like "forecast" in the title or abstract actually report forecasts—but many do (Lewis et al. 2022). Ecologists have now published enough forecasts of enough ecological variables that we can do quantitative reviews of their accuracy (Lewis et al. 2022).

It might be difficult for more-junior readers to appreciate just how big a change this is. In 1991, a few years before I started graduate school, Robert Peters published *A Critique for Ecology*. In it, he calls for ecologists to make what we'd now call forecasts (Peters used the term "predictions"). Peters argues that a scientific field doesn't deserve the moniker "scientific" unless it routinely makes and evaluates forecasts, because making accurate forecasts is the only way to really prove that you know how nature works. *A Critique for Ecology* was very controversial, for various reasons (Lawton 1992; Keddy 1992; Shapiro 1993). But it wasn't widely influential—or at least, it didn't seem like it at the time. I recall that ecologists all talked about Peters's book. And then we all gave a metaphorical shrug (or in a few cases, a metaphorical eyeroll),

Figure 10.1. Proportion of forecasting papers in ecology journals, by publication year

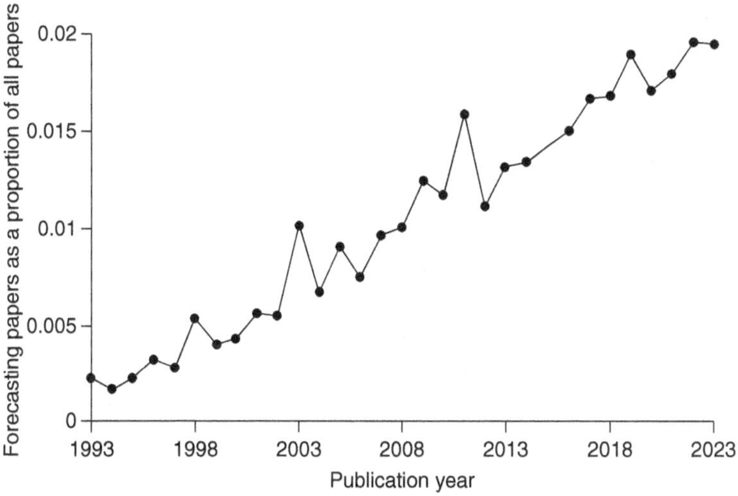

"Forecasting" papers are those with "forecast*" in the title and/or abstract in a Web of Science search, June 2024.

and went back to doing the same ecological research we'd been doing before . . . or did we? Perhaps it's just a just a coincidence that the time series in figure 10.1 started going up a few years after *A Critique for Ecology* was published. It's hard to say; perhaps the then-exploding concern with global climate change is what prompted ecologists to start trying their hand at forecasting. But whatever the reason, something clearly changed. Ecological research has always been about describing nature as it is. But now we're witnessing something almost totally new: a critical mass of ecological research that forecasts how nature will be. I don't know if it will succeed or not (there are reasons to think it won't; see Silver 2015). But I'm excited to find out.

New Approach: Coordinated Distributed Experiments

Ecologists today aren't just pursuing the new research goal of forecasting—they're also inventing new research approaches. One of the most exciting and promising new approaches is known as a coordinated distributed experiment.

A coordinated distributed experiment consists of methodologically identical experiments run simultaneously by different investigators at different locations (H. Fraser et al. 2013; Borer, Harpole, et al. 2014). "Experiment" here means a study using one or more deliberate manipulations, rather than relying on naturally occurring variation. "Coordinated" means that the investigators coordinate with one another so that they all use the same study design and methods.

Perhaps the most influential (although not the first) coordinated distributed experiment in ecology is the Nutrient Network (NutNet) experiment (Borer, Harpole, et al. 2014). NutNet now includes more than 170 sites in grasslands on six continents. At each site, the NutNet experiment crosses a fertilization treatment with a vertebrate herbivore exclusion treatment. It looks at the effects of those manipulations on primary productivity, plant species richness, and various other variables.

Coordinated distributed experiments complement other research approaches in some respects, and improve on them in others (Borer, Harpole, et al. 2014). They complement long-term monitoring networks, which provide observational data, because experimental data is particularly helpful for causal inference and hypothesis testing. They don't replace single-site experiments, because not every worthwhile experiment can be replicated globally. By the same token, they don't replace meta-analyses, because meta-analyses can address many more scientific questions than coordinated distributed experiments can. But coordinated distributed experiments do have some advantages over meta-analyses addressing the same scientific questions. Using the same study design and methods at all sites can sometimes reduce among-site variation in the results (Bebout and Fox 2024). This makes the remaining among-site variation more interpretable by ensuring that it arises for ecological rather than methodological reasons. More importantly, using the same study design and methods at all sites ensures that key predictor variables are measured at all sites in the same manner. Finally, coordinated distributed experiments can be planned so as to include the full range of variation in key ecological variables, rather than having to rely on independently conducted studies that are likely to comprise a statistically biased sample of the ranges of key ecological variables.

I also appreciate coordinated distributed experiments for purely subjective aesthetic reasons. I like the way they mix older and newer research approaches in ecology. On the one hand, they're field experiments, with simple designs, based of necessity on inexpensive, widely available low-tech equipment such as fencing and fertilizer. Volunteer collaborators can't replicate an experiment all over the world if it requires an elaborate study design or expensive high-tech gear. On the other hand, coordinating large numbers of collaborators around the world, and compiling and storing their data, requires making effective use of newer technologies: the internet, relational databases, videoconferencing software, and the like. And it requires developing new ways of working so that hundreds of collaborators can contribute ideas, receive appropriate credit, feel like valued members of the team, and grow into leadership roles. Borer, MacDougall, et al. (2023) discusses these new ways of working in the context of NutNet.

One promising but as yet underused application of coordinated distributed experiments is as a way to facilitate independently conducted studies. Each NutNet site reserves a subset of the plots for site-specific experiments that piggyback on the main NutNet experiment. Each site's researchers can choose their own site-specific questions to address with their own site-specific experiments. Those creative site-specific research projects gain power and generality because of the broader context provided by other NutNet sites. Some site-specific projects also can be scaled up and replicated at other NutNet sites, giving NutNet the ability to address questions that its founders didn't originally envision. There need be no trade-off between centrally coordinated collaborative research and creative independent research by diverse individual investigators. NutNet shows how the two complement one another. One can imagine a future for ecological research in which it becomes standard practice for individual researchers to both pursue their own research programs and participate in one or more coordinated distributed experiments. That's a pretty attractive vision of the future, I think.

I'm not the only one who thinks so. Back in 2019, I polled *Dynamic Ecology* readers about their views on various ways to achieve generality or synthesis in ecological research (J. Fox 2019a). The poll included the approaches to generality discussed in chapter 7, plus a few others, including coordinated distributed experiments. Respondents were asked if they used each approach in their own work, and how important they thought it was for ecologists as a whole to use each approach.

Not surprisingly, ecologists think the most important approaches for ecologists as a whole to use are (i) the approaches they already use themselves, and (ii) the approaches many other ecologists already use (fig. 10.2). With one exception: only about one-quarter of respondents participated in coordinated distributed experiments themselves, but a large majority thought that coordinated distributed experiments were important for ecologists as a whole to pursue (fig. 10.2).

The Hedgehog and the Fox

And that's it. If you're looking for silver bullets, that's as close as I can come. Ecologists should do more forecasting and more coordinated distributed experiments. "More forecasting" and "more coordinated

Figure 10.2. Even ecologists who don't participate in coordinated distributed experiments nevertheless think it is very important for other ecologists to do so

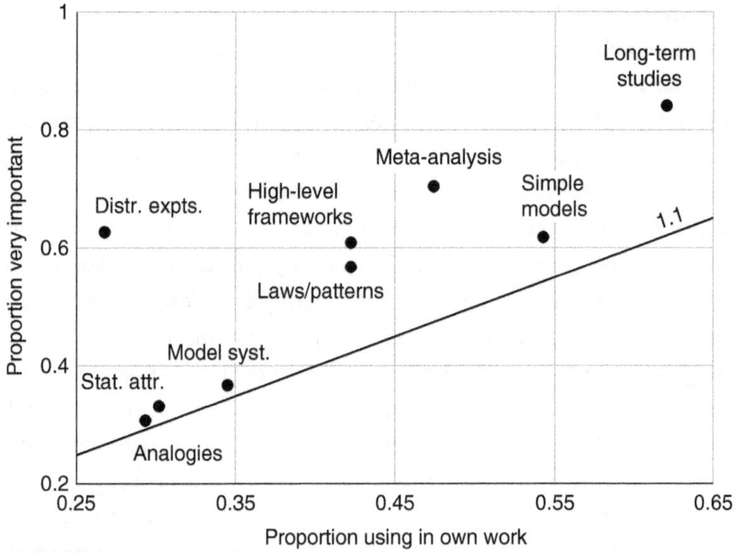

A poll on the *Dynamic Ecology* blog asked respondents which approaches to generality in ecology they use themselves, and which approaches they consider "very important" for the field of ecology as a whole to use. Coordinated distributed experiments stand out as the only approach to generality widely viewed as "very important" for the field of ecology as a whole, despite being used by only a small minority of poll respondents.

Source: J. Fox (2019a).

distributed experiments" are both nice ideas. But neither really sounds like a huge, exciting answer to the question "How can ecologists better harness their diverse ideas, goals, and approaches in the future?"

In my own defense, I don't know that any other ecologist could do much better. And if they could, it would undermine the whole point of this book. This is a book about harnessing the diversity of ideas, goals, and approaches in ecology. If any one ecologist knew enough about all of ecology to explain in detail how ecological research—all of it—ought to be done in the future, that would mean ecologists' ideas and approaches weren't diverse in the first place.

So I don't have a silver bullet. I don't have all the answers, or even most of the answers. What I do have is a vision for how all ecologists, together, can figure out the answers. It's a vision about hedgehogs and foxes.

Back when I founded the *Dynamic Ecology* blog, I gave it a pretentious Latin motto: *Multa novit vulpes*. It translates as "The fox knows many things." That phrase in turn is an English translation of a passage from the Greek poet Archilochus, which contrasts foxes and hedgehogs:

> The fox knows many things, but the hedgehog knows one big thing.

Like many people, I first encountered that line in intellectual historian Isaiah Berlin's 1953 essay *The Hedgehog and the Fox*. Berlin (1953, 1–2) interprets Archilochus as demarcating two kinds of writers and thinkers:

> There is a line among the fragments of the Greek poet Archilochus which says: "The fox knows many things, but the hedgehog knows one big thing." Scholars have differed about the correct interpretation of these dark words, which may mean no more than that the fox, for all his cunning, is defeated by the hedgehog's one defence. But, taken figuratively, the words can be made to yield a sense in which they mark one of the deepest differences which divide writers and thinkers, and, it may be, human beings in general. For there exists a great chasm between those, on one side, who relate everything to a single central vision, one system, less or more coherent or articulate, in terms of which they understand, think and feel—a single, universal, organising principle in terms of which alone all that they are

> and say has significance—and, on the other side, those who pursue many ends, often unrelated and even contradictory, connected, if at all, only in some de facto way, for some psychological or physiological cause, related to no moral or aesthetic principle. These last lead lives, perform acts and entertain ideas that are centrifugal rather than centripetal; their thought is scattered or diffused, moving on many levels, seizing upon the essence of a vast variety of experiences and objects for what they are in themselves, without, consciously or unconsciously, seeking to fit them into, or exclude them from, any one unchanging, all-embracing, sometimes self-contradictory and incomplete, at times fanatical, unitary inner vision. The first kind of intellectual and artistic personality belongs to the hedgehogs, the second to the foxes.

In one of my first blog posts for *Dynamic Ecology*, I mused on whether I am a fox in name only (J. Fox 2012c). I like to think of myself as a fox in Berlin's sense: a generalist who knows and thinks about many different topics. I hope I've written a book that reflects and conveys my own foxy self-image. But in some important ways, I am very much a hedgehog—or a one-trick pony, as they're also known. After all, I've spent my entire professional life in the ivory tower of academia, and my entire faculty career at a single university. I've been working in the same study system—protist microcosms—since I was an undergraduate. I've been blogging for more than 10 years without ever taking up other forms of online communication, such as social media. And although *I* think this book uses a pretty wide range of examples to make its points, no doubt the choice of examples retains traces of my own biases. I can certainly imagine that some readers will find my choices of examples disappointingly narrow, and not without reason.

But the same could be said of any ecologist. We're all metaphorical hedgehogs in many respects. Indeed, it's tempting to think that we *have* to be. How could we be anything but hedgehogs, given the size and growth rate of the field of ecology? Surely the only way to keep up is to specialize, right? To know "one big thing," as it were.

The hope is that it's OK if we all specialize in our own narrow lines of research, because a bunch of hedgehogs adds up to one fox. That if every ecologist knows "one big thing," and different ecologists know different big things, then collectively we know "many things." Richard Levins articulated that hope way back in 1966. He was talking about model building in population ecology, but he equally well could've been talking about the field of ecology writ large:

> The multiplicity of models is imposed by the contradictory demands of a complex, heterogeneous nature and a mind that can only cope with a few variables at a time; by the contradictory desiderata of generality, realism, and precision; by the need to understand and also to control; even by the opposing esthetic standards which emphasize the stark simplicity and power of a general theorem as against the richness and diversity of living nature. These conflicts are irreconcilable. Therefore, the alternative approaches even of contending schools are part of a larger mixed strategy. (Levins 1966, 431)

Levins was right, but he didn't go quite far enough. As described in chapters 1 and 7, ecologists are starting to run up against the limits of how much we can learn if everyone just does their own small-sample empirical studies (collaborative or otherwise), and then uses meta-analysis to synthesize those studies. If each hedgehog doesn't actually know "one big thing" (because they're all doing low-powered studies of noisy, heterogeneous phenomena), then they add up to a fox that doesn't know much of anything.

Maybe—probably—you think that's too pessimistic, even cynical. Maybe you think every one of our ecologist hedgehogs does know (at least) "one big thing." If so, fair enough—you might be right. But even if every hedgehog *does* know "one big thing," well, then the field of ecology as a whole knows only what all the researchers composing it know. The field is the sum of its parts—but no greater than that. It's foxiness of a sort, but a pretty minimal foxiness.

Ecologists can do better than that, but first we have to want to. We have to believe that, as a field, we can be *more* than a bunch of hedgehogs. We have to believe that ecology as a field can be more than just "the sum of all the things known by all the people who call themselves 'ecologists.'" We can't just be satisfied with the status quo. We can't just be satisfied with chipping away at scientific questions we never answer beyond saying, "It depends, and we don't really know on what." We have to start treating the noisiness and heterogeneity of the phenomena we study as a challenge to be overcome, not as an ineffability to be celebrated or as an all-purpose excuse for spinning our collective wheels. We have to be dissatisfied enough with where our field stands right now to want it to change. We have to be ambitious enough to believe we can change it. And we have to be honest enough with ourselves to realize it's not going to change automatically, as a happy by-product of the increasing diversity of ecologists. Ecologists are increasingly diverse in various dimensions, and that's great. But the diversity of our ideas, goals, and approaches hasn't kept up. Nor has our ability to harness that diversity through selection effects, complementarity effects, and the use of different intellectual tools for different jobs.

So if we want the field of ecology to be *greater* than the sum of its parts, we can't all just be hedgehogs. Each of us has to become at least a bit of a fox—a bit of a generalist. We have to be curious about, and care about, ecological research outside our own narrow subfield, and find creative ways for that research to shape our own. We have to find new ways to draw strength from ecological research that asks different questions than ours, using different methods, with different goals. We have to realize that the challenges to research progress in our own narrow subfield have analogs in other subfields, and even in fields outside ecology. We have to believe we can learn from those analogs. We have to recognize that ecologists' struggles to select for the right things aren't idiosyncratic—those struggles run to type. We have to . . .

OK, I'll stop. You get the point. A bunch of hedgehogs doesn't add up to much of a fox—but a bunch of foxes adds up to one much smarter fox.

I know that ecologists can change, because they've done it before. Starting in the late 1980s, ecologists responded to public recognition of global climate change by revolutionizing the questions we ask. Not too long after, we took advantage of rapid advances in computer networking and software to revolutionize the statistics we use, the geographic scales at which we collaborate, and the use we make of existing data. These weren't the first revolutions in ecology, and they shouldn't be the last. But the next revolution won't be prompted by external events or new technologies. It's going to have to come from us.

I hope this book makes some small contribution to making the next revolution happen. You can see the seeds of it already. You see it in exemplary studies that integrate complementary lines of evidence, which more ecologists could emulate. You see it in new research approaches like coordinated distributed experiments, which were born out of a recognition of the limitations of existing research approaches. You can see it in the recent growth of interest in forecasting, historically a rare research goal in ecology. You can see it in other recent ecology books, which share my own sense that ecology is ripe for a rethink (Cousens 2023). And most importantly, I'm sure you can see it in a bunch of places I'm totally unaware of. I don't know which of those diverse seeds will germinate and bear fruit—which ones we'll look back on 50 years from now while saying, "There's where the revolution started." But that's OK; the important thing is that the seeds are there. As science fiction author William Gibson purportedly said:

> The future has arrived—it's just not evenly distributed yet.

Acknowledgments

Christie Henry originally approached me about writing a book for the University of Chicago Press. I will be forever grateful to her for giving me an opportunity I hadn't sought, and for continuing to support me as my vision for the book evolved over time. Thank you as well to Joseph Calamia, who took over as my editor after Christie Henry moved to a new position and ably shepherded the book to completion.

Thank you to Tina Harrison, Meghan Duffy, Brian McGill, Peter Adler, Mark Vellend, Colin Kremer, Eric Charnov, Robin Snyder, Stephen Heard, Jeff Houlahan, Rachel Germain and her lab group, and an anonymous reviewer for feedback on draft chapters. This book would've been much worse without your input. The faults that no doubt remain are solely my own responsibility.

A previous version of chapter 7 was originally published in the Spring 2019 issue of *Philosophical Topics*. Thank you to Jay Odenbaugh for inviting me to write something for that issue, thereby allowing me to fulfill a longtime dream of pretending to be a philosopher. That chapter also benefited from feedback from my colleague Marc Ereshefsky and the philosophy graduate students at the University of Calgary.

Many of the ideas for this book grew out of my blogging, first for *Oikos Blog* and then for the blog I founded, *Dynamic Ecology*. Thank you to Chris Lortie for starting *Oikos Blog*; otherwise, I'd likely have stuck to reading blogs rather than writing them. And although I founded *Dynamic Ecology*, it isn't just my blog. To my blogging colleagues Meghan

Duffy and Brian McGill: thank you so, so much for joining me. You made *Dynamic Ecology* much better than I ever could have on my own. I couldn't have asked for better collaborators, or better friends. Thanks as well to everyone who has read and commented on *Dynamic Ecology* over the years. This book wouldn't exist without your encouragement and enthusiasm.

References

AAAS American Association for the Advancement of Science. n.d. "Historical Trends in Federal R&D." Accessed October 24, 2024. https://www.aaas.org/programs/r-d-budget-and-policy/historical-trends-federal-rd.

Aarssen, Lonnie W. 1997. "High Productivity in Grassland Ecosystems: Effected by Species Diversity or Productive Species?" *Oikos* 80 (1): 183–84. https://doi.org/10.2307/3546531.

Abrams, Peter A. 1993. "Effect of Increased Productivity on the Abundances of Trophic Levels." *American Naturalist* 141 (3): 351–71. https://doi.org/10.1086/285478.

———. 2015. "Why Ratio Dependence Is (Still) a Bad Model of Predation." *Biological Reviews of the Cambridge Philosophical Society* 90 (3): 794–814. https://doi.org/10.1111/brv.12134.

Allesina, Stefano, and Si Tang. 2012. "Stability Criteria for Complex Ecosystems." *Nature* 483 (7388): 205–8. https://doi.org/10.1038/nature10832.

Alroy, John. 2015. "The Shape of Terrestrial Abundance Distributions." *Science Advances* 1 (8): e1500082. https://doi.org/10.1126/sciadv.1500082.

Anderson, Sean C., Paul R. Elsen, Brent B. Hughes, Rebecca K. Tonietto, Molly C. Bletz, David A. Gill, Meredith A. Holgerson, et al. 2021. "Trends in Ecology and Conservation over Eight Decades." *Frontiers in Ecology and the Environment* 19 (5): 274–82. https://doi.org/10.1002/fee.2320.

Archer, C. R., C. W. W. Pirk, L. G. Carvalheiro, and S. W. Nicolson. 2014. "Economic and Ecological Implications of Geographic Bias in Pollinator Ecology in the Light of Pollinator Declines." *Oikos* 123 (4): 401–7. https://doi.org/10.1111/j.1600-0706.2013.00949.x.

Arditti, Joseph, John Elliott, Ian J. Kitching, and Lutz T. Wasserthal. 2012. "'Good Heavens What Insect Can Suck It'—Charles Darwin, *Angraecum sesquipedale* and *Xanthopan morganii praedicta*." *Botanical Journal of the Linnean Society* 169 (3): 403–32. https://doi.org/10.1111/j.1095-8339.2012.01250.x.

Ashworth, Scott, Christopher R. Berry, and Ethan Bueno de Mesquita. 2021. *Theory and Credibility: Integrating Theoretical and Empirical Social Science*. Princeton University Press.

Bakker, Lisette M., Kathryn E. Barry, Liesje Mommer, and Jasper van Ruijven. 2021. "Focusing on Individual Plants to Understand Community Scale Biodiversity Effects: The Case of Root Distribution in Grasslands." *Oikos* 130 (11): 1954–66. https://doi.org/10.1111/oik.08113.

Baldridge, Elita, David J. Harris, Xiao Xiao, and Ethan P. White. 2016. "An Extensive Comparison of Species-Abundance Distribution Models." *PeerJ* 4 (December): e2823. https://doi.org/10.7717/peerj.2823.

Banavar, Jayanth R., Melanie E. Moses, James H. Brown, John Damuth, Andrea Rinaldo, Richard M. Sibly, and Amos Maritan. 2010. "A General Basis for Quarter-Power Scaling in Animals." *Proceedings of the National Academy of Sciences* 107 (36): 15816–20. https://doi.org/10.1073/pnas.1009974107.

Barbier, Matthieu, Guy Bunin, and Mathew A. Leibold. 2025. "Getting More by Asking for Less: Linking Species Interactions to Species Co-distributions in Metacommunities." *Peer Community Journal* 5: e1. https://doi.org/10.24072/pcjournal.483.

Barner, Allison K., Kyle E. Coblentz, Sally D. Hacker, and Bruce A. Menge. 2018. "Fundamental Contradictions among Observational and Experimental Estimates of Non-trophic Species Interactions." *Ecology* 99 (3): 557–66. https://doi.org/10.1002/ecy.2133.

Barraquand, Frédéric. 2014. "Functional Responses and Predator-Prey Models: A Critique of Ratio Dependence." *Theoretical Ecology* 7 (1): 3–20. https://doi.org/10.1007/s12080-013-0201-9.

Barraquand, Frédéric, Stilianos Louca, Karen C. Abbott, Christina A. Cobbold, Flora Cordoleani, Donald L. DeAngelis, Bret D. Elderd, et al. 2017. "Moving Forward in Circles: Challenges and Opportunities in Modelling Population Cycles." *Ecology Letters* 20 (8): 1074–92. https://doi.org/10.1111/ele.12789.

Baselga, Andrés, and Fabien Leprieur. 2015. "Comparing Methods to Separate Components of Beta Diversity." *Methods in Ecology and Evolution* 6 (9): 1069–79. https://doi.org/10.1111/2041-210X.12388.

Bebout, Julia, and Jeremy W. Fox. 2024. "Coordinated Distributed Experiments in Ecology Do Not Consistently Reduce Heterogeneity in Effect Size." *Oikos* 2024 (6): e10722. https://doi.org/10.1111/oik.10722.

Beck, Christopher, Kate Boersma, C. Susannah Tysor, and George Middendorf. 2014. "Diversity at 100: Women and Underrepresented Minorities in the ESA." *Frontiers in Ecology and the Environment* 12 (8): 434–36. https://doi.org/10.1890/14.WB.011.

Begon, M., J. L. Harper, and C. R. Townsend. 1990. *Ecology: Individuals, Populations and Communities*. 2nd ed. Wiley & Sons.

Bell, Graham. 2001. "Neutral Macroecology." *Science* 293 (5539): 2413–18. https://doi.org/10.1126/science.293.5539.2413.

Belluz, Julia. 2016. "The 7 Biggest Problems Facing Science, According to 270 Scientists." *Vox*, July 14. https://www.vox.com/2016/7/14/12016710/science-challeges-research-funding-peer-review-process.

Berlin, Isaiah. 1953. *The Hedgehog and the Fox: An Essay on Tolstoy's View of History*. Weidenfeld & Nicolson.

Berry, Andrew, and Janet Browne. 2022. "Mendel and Darwin." *Proceedings of the National Academy of Sciences* 119 (30): e2122144119. https://doi.org/10.1073/pnas.2122144119.

Berryman, Alan A. 2003. "On Principles, Laws and Theory in Population Ecology." *Oikos* 103 (3): 695–701. https://doi.org/10.1034/j.1600-0706.2003.12810.x.

Betini, Gustavo S., Tal Avgar, and John M. Fryxell. 2017. "Why Are We Not Evaluating Multiple Competing Hypotheses in Ecology and Evolution?" *Royal Society Open Science* 4 (1): 160756. https://doi.org/10.1098/rsos.160756.

Betts, Matthew G., Adam S. Hadley, David W. Frey, Sarah J. K. Frey, Dusty Gannon, Scott H. Harris, Hankyu Kim, et al. 2021. "When Are Hypotheses Useful in Ecology and Evolution?" *Ecology and Evolution* 11 (11): 5762–76. https://doi.org/10.1002/ece3.7365.

Bishop, Jacob, and Shinichi Nakagawa. 2021. "Quantifying Crop Pollinator Dependence and Its Heterogeneity Using Multi-level Meta-analysis." *Journal of Applied Ecology* 58 (5): 1030–42. https://doi.org/10.1111/1365-2664.13830.

Bjørnstad, Ottar N., and Bryan T. Grenfell. 2001. "Noisy Clockwork: Time Series Analysis of Population Fluctuations in Animals." *Science* 293 (5530): 638–43. https://doi.org/10.1126/science.1062226.

Blowes, Shane A., Sarah R. Supp, Laura H. Antão, Amanda Bates, Helge Bruelheide, Jonathan M. Chase, Faye Moyes, et al. 2019. "The Geography of Biodiversity Change in Marine and Terrestrial Assemblages." *Science* 366 (6463): 339–45. https://doi.org/10.1126/science.aaw1620.

Bolker, Ben. 2005. "Other People's Data." *BioScience* 55 (7): 550–51.

Boltovskoy, Demetrio, Nancy M. Correa, Lyubov E. Burlakova, Alexander Y. Karatayev, Erik V. Thuesen, Francisco Sylvester, and Esteban M. Paolucci. 2021. "Traits and Impacts of Introduced Species: A Quantitative Review of Meta-analyses." *Hydrobiologia* 848 (9): 2225–58. https://doi.org/10.1007/s10750-020-04378-9.

Borer, Elizabeth T., W. Stanley Harpole, Peter B. Adler, Eric M. Lind, John L. Orrock, Eric W. Seabloom, and Melinda D. Smith. 2014. "Finding Generality in Ecology: A Model for Globally Distributed Experiments." *Methods in Ecology and Evolution* 5 (1): 65–73. https://doi.org/10.1111/2041-210X.12125.

Borer, Elizabeth T., Andrew S. MacDougall, Carly J. Stevens, Lauren L. Sullivan, Peter A. Wilfahrt, and Eric W. Seabloom. 2023. "Writing a Massively Multi-authored Paper: Overcoming Barriers to Meaningful Authorship for All." *Methods in Ecology and Evolution* 14 (6): 1432–42. https://doi.org/10.1111/2041-210X.14096.

Brandell, Ellen E., Daniel J. Becker, Laura Sampson, and Kristian M. Forbes. 2021. "Demography, Education, and Research Trends in the Interdisciplinary Field of Disease Ecology." *Ecology and Evolution* 11 (24): 17581–92. https://doi.org/10.1002/ece3.8466.

Breznau, Nate, Eike Mark Rinke, Alexander Wuttke, Hung H. V. Nguyen, Muna Adem, Jule Adriaans, Amalia Alvarez-Benjumea, et al. 2022. "Observing Many Researchers Using the Same Data and Hypothesis Reveals a Hidden Universe of Uncertainty." *Proceedings of the National Academy of Sciences* 119 (44): e2203150119. https://doi.org/10.1073/pnas.2203150119.

Brillinger, David R. 2002. "John W. Tukey: His Life and Professional Contributions." *Annals of Statistics* 30 (6): 1535–75. https://doi.org/10.1214/aos/1043351246.

Brisson, Jacques, Mariana Rodriguez, Charles A. Martin, and Raphaël Proulx. 2020. "Plant Diversity Effect on Water Quality in Wetlands: A Meta-analysis Based on Experimental Systems." *Ecological Applications* 30 (4): e02074. https://doi.org/10.1002/eap.2074.

Browman, Howard I. 2017. "Towards a Broader Perspective on Ocean Acidification Research." *ICES Journal of Marine Science* 74 (4): 889–94. https://doi.org/10.1093/icesjms/fsx073.

Brown, Bryan, Eric Sokol, James Skelton, and Brett Tornwall. 2017. "Making Sense of Metacommunities: Dispelling the Mythology of a Metacommunity Typology." *Oecologia* 183 (March). https://doi.org/10.1007/s00442-016-3792-1.

Brown, James H. 1995. *Macroecology*. University of Chicago Press.

———. 1997. "EcoEssay Series Number 1." National Center for Ecological Analysis and Synthesis. https://www.nceas.ucsb.edu/ecoessays/brown.

Brown, James H., and Mark V. Lomolino. 2000. "Concluding Remarks: Historical Perspective and the Future of Island Biogeography Theory." *Global Ecology and Biogeography* 9 (1): 87–92.

Brown, James H., and Brian A. Maurer. 1989. "Macroecology: The Division of Food and Space among Species on Continents." *Science* 243 (4895): 1145–50.

Bujang, Mohamad Adam, and Nurakmal Baharum. 2016. "Sample Size Guideline for Correlation Analysis." *World Journal of Social Science Research* 3 (1): 37. https://doi.org/10.22158/wjssr.v3n1p37.

Burgess, Benjamin J., Michelle C. Jackson, and David J. Murrell. 2022. "Are Experiment Sample Sizes Adequate to Detect Biologically Important

Interactions between Multiple Stressors?" *Ecology and Evolution* 12 (9): e9289. https://doi.org/10.1002/ece3.9289.

Burnham, Kenneth P., and David R. Anderson. 2003. *Model Selection and Multimodel Inference: A Practical Information-Theoretic Approach*. Springer Science & Business Media.

Burns, C. Sean, and Charles W. Fox. 2017. "Language and Socioeconomics Predict Geographic Variation in Peer Review Outcomes at an Ecology Journal." *Scientometrics* 113 (2): 1113–27. https://doi.org/10.1007/s11192-017-2517-5.

Butterfield, Jeremy. 2019. "Lost in Math? A Review of 'Lost in Math: How Beauty Leads Physics Astray,' by Sabine Hossenfelder." Preprint, arXiv, February 9. https://doi.org/10.48550/arXiv.1902.03480.

Campbell, Rebecca, Rachael Goodman-Williams, Hannah Feeney, and Giannina Fehler-Cabral. 2018. "Assessing Triangulation Across Methodologies, Methods, and Stakeholder Groups: The Joys, Woes, and Politics of Interpreting Convergent and Divergent Data." *American Journal of Evaluation* 41 (November): 109821401880419. https://doi.org/10.1177/1098214018804195.

Campos-Arceiz, Ahimsa, Richard B. Primack, and Lian Pin Koh. 2015. "Reviewer Recommendations and Editors' Decisions for a Conservation Journal: Is It Just a Crapshoot? And Do Chinese Authors Get a Fair Shot?" *Biological Conservation* 186 (June): 22–27. https://doi.org/10.1016/j.biocon.2015.02.025.

Cantrell, Robert Stephen, and Chris Cosner. 2004. *Spatial Ecology via Reaction-Diffusion Equations*. John Wiley & Sons.

Cardinale, Bradley. 2014. "Overlooked Local Biodiversity Loss." *Science* 344 (6188): 1098. https://doi.org/10.1126/science.344.6188.1098-a.

Carpenter, Stephen R. 1996. "Microcosm Experiments Have Limited Relevance for Community and Ecosystem Ecology." *Ecology* 77 (3): 677–80. https://doi.org/10.2307/2265490.

Carroll, Sean. 2006. "The Trouble with Physics." October 3. https://www.preposterousuniverse.com/blog/2006/10/03/the-trouble-with-physics/.

Case, T. J. 1990. "Invasion Resistance Arises in Strongly Interacting Species-Rich Model Competition Communities." *Proceedings of the National Academy of Sciences* 87 (24): 9610–14. https://doi.org/10.1073/pnas.87.24.9610.

Castagneyrol, Bastien, Baptiste Bedessem, and Romain Julliard. 2023. "Is Ecology Different When Studied with Citizen Scientists? A Bibliometric Analysis." *Ecology and Evolution* 13 (9). https://doi.org/10.1002/ece3.10488.

Caswell, Hal. 1988. "Theory and Models in Ecology: A Different Perspective." *Ecological Modelling* 43 (1): 33–44. https://doi.org/10.1016/0304-3800(88)90071-3.

Cattin, Marie-France, Louis-Félix Bersier, Carolin Banašek-Richter, Richard Baltensperger, and Jean-Pierre Gabriel. 2004. "Phylogenetic Constraints and Adaptation Explain Food-Web Structure." *Nature* 427 (6977): 835–39. https://doi.org/10.1038/nature02327.

Charney, Davida. 1993. "A Study in Rhetorical Reading: How Evolutionists Read 'The Spandrels of San Marco.'" In *Understanding Scientific Prose*, edited by Jack Selzer, 203–31. University of Wisconsin Press.

Charnov, Eric L. 1976. "Optimal Foraging, the Marginal Value Theorem." *Theoretical Population Biology* 9 (2): 129–36. https://doi.org/10.1016/0040-5809(76)90040-X.

Charnov, Eric, and James Gillooly. 2003. "Thermal Time: Body Size, Food Quality and the 10C Rule." *Biology Faculty & Staff Publications* 5 (January): 43–51.

Chesson, Peter. 2000. "Mechanisms of Maintenance of Species Diversity." *Annual Review of Ecology and Systematics* 31 (1): 343–66. https://doi.org/10.1146/annurev.ecolsys.31.1.343.

Chesson, Peter, and Nancy Huntly. 1997. "The Roles of Harsh and Fluctuating Conditions in the Dynamics of Ecological Communities." *American Naturalist* 150 (5): 519–53. https://doi.org/10.1086/286080.

Cid, Carmen R, and Gillian Bowser. 2015. "Breaking Down the Barriers to Diversity in Ecology." *Frontiers in Ecology and the Environment* 13 (4): 179. https://doi.org/10.1890/1540-9295-13.4.179.

Clark, Brady. 2010. "Evolutionary Frameworks for Language Change: The Price Equation Approach." *Language and Linguistics Compass* 4 (6): 363–76. https://doi.org/10.1111/j.1749-818X.2010.00200.x.

Clark, Michael. n.d. "R and Social Science." Accessed October 24, 2024. https://m-clark.github.io/docs/RSocialScience.pdf.

Clark, Timothy D., Graham D. Raby, Dominique G. Roche, Sandra A. Binning, Ben Speers-Roesch, Fredrik Jutfelt, and Josefin Sundin. 2020. "Ocean Acidification Does Not Impair the Behaviour of Coral Reef Fishes." *Nature* 577 (7790): 370–75. https://doi.org/10.1038/s41586-019-1903-y.

Cleasby, Ian R., Barbara J. Morrissey, Mark Bolton, Ellie Owen, Linda Wilson, Saskia Wischnewski, and Shinichi Nakagawa. 2021. "What Is Our Power to Detect Device Effects in Animal Tracking Studies?" *Methods in Ecology and Evolution* 12 (7): 1174–85. https://doi.org/10.1111/2041-210X.13598.

Clements, Jeff C., Josefin Sundin, Timothy D. Clark, and Fredrik Jutfelt. 2022. "Meta-analysis Reveals an Extreme 'Decline Effect' in the Impacts of Ocean Acidification on Fish Behavior." *PLOS Biology* 20 (2): e3001511. https://doi.org/10.1371/journal.pbio.3001511.

Cohen, Dan. 1966. "Optimizing Reproduction in a Randomly Varying Environment." *Journal of Theoretical Biology* 12 (1): 119–29. https://doi.org/10.1016/0022-5193(66)90188-3.

Cohen, Joel E., Tomas Jonsson, and Stephen R. Carpenter. 2003. "Ecological Community Description Using the Food Web, Species Abundance, and

Body Size." *Proceedings of the National Academy of Sciences* 100 (4): 1781–86. https://doi.org/10.1073/pnas.232715699.

Cohen, Joel E., and C. M. Newman. 1985. "A Stochastic Theory of Community Food Webs I. Models and Aggregated Data." *Proceedings of the Royal Society B: Biological Sciences* 224 (1237): 421–48. https://doi.org/10.1098/rspb.1985.0042.

Colautti, Robert I., and Hugh J. MacIsaac. 2004. "A Neutral Terminology to Define 'Invasive' Species." *Diversity and Distributions* 10 (2): 135–41. https://doi.org/10.1111/j.1366-9516.2004.00061.x.

Collins, Sinéad, and Andy Gardner. 2009. "Integrating Physiological, Ecological and Evolutionary Change: A Price Equation Approach." *Ecology Letters* 12 (8): 744–57. https://doi.org/10.1111/j.1461-0248.2009.01340.x.

Colwell, Robert K., and George C. Hurtt. 1994. "Nonbiological Gradients in Species Richness and a Spurious Rapoport Effect." *American Naturalist* 144 (4): 570–95.

Colwell, Robert K., and D. C. Lees. 2000. "The Mid-domain Effect: Geometric Constraints on the Geography of Species Richness." *Trends in Ecology & Evolution* 15 (2): 70–76. https://doi.org/10.1016/s0169-5347(99)01767-x.

Colwell, Robert K., C. Rahbek, and N. J. Gotelli. 2005. "The Mid-domain Effect: There's a Baby in the Bathwater." *American Naturalist* 166 (5): E149–54. https://doi.org/10.1086/491689.

Colwell, Robert K., and D. W. Winkler. 1984. "A Null Model for Null Models in Biogeography." In *Ecological Communities: Conceptual Issues and the Evidence*, edited by D. R. Strong Jr., D. Simberloff, L. G. Abele, and A. B. Thistle, 344–59. Princeton University Press.

Colyvan, M., and L. R. Ginzburg. 2003. "Laws of Nature and Laws of Ecology." *Oikos* 101 (3): 649–53. https://doi.org/10.1034/j.1600-0706.2003.12349.x.

Connolly, Sean R. 2005. "Process-Based Models of Species Distributions and the Mid-domain Effect." *American Naturalist* 166 (1): 1–11. https://doi.org/10.1086/430638.

Connor, Edward F., and Daniel Simberloff. 1979. "The Assembly of Species Communities: Chance or Competition?" *Ecology* 60 (6): 1132–40. https://doi.org/10.2307/1936961.

Cooper, Gregory J. 2003. *The Science of the Struggle for Existence: On the Foundations of Ecology*. Cambridge Studies in Philosophy and Biology. Cambridge University Press. https://doi.org/10.1017/CBO9780511720154.

Cornell, Howard V. 1985. "Local and Regional Richness of Cynipine Gall Wasps on California Oaks." *Ecology* 66 (4): 1247–60. https://doi.org/10.2307/1939178.

Cornell, Howard V., and J. H. Lawton. 1992. "Species Interactions, Local and Regional Processes, and Limits to the Richness of Ecological Communities:

A Theoretical Perspective." *Journal of Animal Ecology* 61 (1): 1–12. https://doi.org/10.2307/5503.

Cosentino, Francesca, and Luigi Maiorano. 2021. "Is Geographic Sampling Bias Representative of Environmental Space?" *Ecological Informatics* 64 (September): 101369. https://doi.org/10.1016/j.ecoinf.2021.101369.

Costa-Pereira, Raul, Remington J. Moll, Brett R. Jesmer, and Walter Jetz. 2022. "Animal Tracking Moves Community Ecology: Opportunities and Challenges." *Journal of Animal Ecology* 91 (7): 1334–44. https://doi.org/10.1111/1365-2656.13698.

Costello, Laura, and Jeremy W. Fox. 2022. "Decline Effects Are Rare in Ecology." *Ecology* 103 (6): e3680. https://doi.org/10.1002/ecy.3680.

Cottenie, Karl. 2005. "Integrating Environmental and Spatial Processes in Ecological Community Dynamics." *Ecology Letters* 8 (11): 1175–82. https://doi.org/10.1111/j.1461-0248.2005.00820.x.

Cottingham, K. L., B. L. Brown, and J. T. Lennon. 2001. "Biodiversity May Regulate the Temporal Variability of Ecological Systems." *Ecology Letters* 4 (1): 72–85. https://doi.org/10.1046/j.1461-0248.2001.00189.x.

Cousens, Roger D. 2023. *Effective Ecology: Seeking Success in a Hard Science*. 1st ed. Boca Raton: CRC Press.

Cox, Kevin R. 2021. "What Happened to 'Geography'?" *Unfashionable Geographies* (blog), May 28. https://kevinrcox.wordpress.com/2021/05/28/what-happened-to-geography/.

Coyne, Jerry A. 1992. "Much Ado about Species." *Nature* 357:289–90.

Coyne, Jerry A., and H. Allen Orr. 2004. *Speciation*. Illus. ed. Sinauer Associates, an imprint of Oxford University Press.

Craven, Dylan, Marten Winter, Konstantin Hotzel, Jitendra Gaikwad, Nico Eisenhauer, Martin Hohmuth, Birgitta König-Ries, and Christian Wirth. 2019. "Evolution of Interdisciplinarity in Biodiversity Science." *Ecology and Evolution* 9 (12): 6744–55. https://doi.org/10.1002/ece3.5244.

Crystal-Ornelas, Robert, and Julie Lockwood. 2020a. "The 'Known Unknowns' of Invasive Species Impact Measurement." *Biological Invasions* 22 (April). https://doi.org/10.1007/s10530-020-02200-0.

———. 2020b. "Cumulative Meta-analysis Identifies Declining but Negative Impacts of Invasive Species on Richness after 20 Yr." *Ecology* 101 (8): e03082. https://doi.org/10.1002/ecy.3082.

Currie, David J., and Jeremy T. Kerr. 2008. "Tests of the Mid-domain Hypothesis: A Review of the Evidence." *Ecological Monographs* 78 (1): 3–18. https://doi.org/10.1890/06-1302.1.

Dallas, Tad, Brett A. Melbourne, and Alan Hastings. 2019. "When Can Competition and Dispersal Lead to Checkerboard Distributions?" *Journal of Animal Ecology* 88 (2): 269–76. https://doi.org/10.1111/1365-2656.12913.

Damuth, John. 1991. "Of Size and Abundance." *Nature* 351 (6324): 268–69. https://doi.org/10.1038/351268a0.

D'Andrea, Rafael, Gabriel Khattar, Thomas Koffel, Veronica Frans, Leonora Bittleston, and Catalina Cuellar-Gempeler. n.d. "Reciprocal Inhibition and Competitive Hierarchy Cause Negative Biodiversity-Ecosystem Function Relationships." Accessed October 28, 2024. https://www.authorea.com/doi/full/10.22541/au.169470011.16685842/v1?commit=0241bf66cad66060ecd414564b1161b1a9ad3ba7.

Darwin, Charles. 1859. *On the Origin of Species by Means of Natural Selection, or, The Preservation of Favoured Races in the Struggle for Life*. London: J. Murray. Available at https://www.loc.gov/resource/rbctos.2017gen17473/.

Davidson, Kate E., Mike S. Fowler, Martin W. Skov, Stefan H. Doerr, Nicola Beaumont, and John N. Griffin. 2017. "Livestock Grazing Alters Multiple Ecosystem Properties and Services in Salt Marshes: A Meta-analysis." *Journal of Applied Ecology* 54 (5): 1395–1405. https://doi.org/10.1111/1365-2664.12892.

Davis, A. J., L. S. Jenkinson, J. H. Lawton, B. Shorrocks, and S. Wood. 1998. "Making Mistakes When Predicting Shifts in Species Range in Response to Global Warming." *Nature* 391 (6669): 783–86. https://doi.org/10.1038/35842.

Davis, Phil. 2016. "PNAS: Tighter Editorial Policy Improves NAS Papers." *The Scholarly Kitchen* (blog), January 14. https://scholarlykitchen.sspnet.org/2016/01/14/pnas-tighter-editorial-policy-improves-nas-papers/.

Davis, Wade. 2021. "Why Anthropology Matters." *Scientific American*, February 21. https://www.scientificamerican.com/article/why-anthropology-matters/.

Dayton, Paul K., and Enric Sala. 2001. "Natural History: The Sense of Wonder, Creativity and Progress in Ecology." *Scientia Marina* 65 (S2): 199–206. https://doi.org/10.3989/scimar.2001.65s2199.

de Ruiter, P. C., A. M. Neutel, and J. C. Moore. 1995. "Energetics, Patterns of Interaction Strengths, and Stability in Real Ecosystems." *Science* 269 (5228): 1257–60. https://doi.org/10.1126/science.269.5228.1257.

Diamond, Jared M. 1975. "Assembly of Species Communities." In *Ecology and Evolution of Communities*, edited by Martin L. Cody and Jared M. Diamond, 342–444. Belknap Press.

Di Marco, Moreno, Sarah Chapman, Glenn Althor, Stephen Kearney, Charles Besancon, Nathalie Butt, Joseph M. Maina, et al. 2017. "Changing Trends and Persisting Biases in Three Decades of Conservation Science." *Global Ecology and Conservation* 10 (April): 32–42. https://doi.org/10.1016/j.gecco.2017.01.008.

Dimson, Monica, Lucas Berio Fortini, Morgan W. Tingley, and Thomas W. Gillespie. 2023. "Citizen Science Can Complement Professional Invasive

Plant Surveys and Improve Estimates of Suitable Habitat." *Diversity and Distributions* 29 (9): 1141–56. https://doi.org/10.1111/ddi.13749.

Dixson, Danielle L., David Abrego, and Mark E. Hay. 2014. "Chemically Mediated Behavior of Recruiting Corals and Fishes: A Tipping Point That May Limit Reef Recovery." *Science* 345 (6199): 892. https://doi.org/10.1126/science.1255057.

Dixson, Danielle L., Philip L. Munday, and Geoffrey P. Jones. 2010. "Ocean Acidification Disrupts the Innate Ability of Fish to Detect Predator Olfactory Cues." *Ecology Letters* 13 (1): 68–75. https://doi.org/10.1111/j.1461-0248.2009.01400.x.

Doak, D. F., D. Bigger, E. K. Harding, M. A. Marvier, R. E. O'Malley, and D. Thomson. 1998. "The Statistical Inevitability of Stability-Diversity Relationships in Community Ecology." *American Naturalist* 151 (3): 264–76. https://doi.org/10.1086/286117.

Dochtermann, Ned A., and Stephen H. Jenkins. 2011. "Developing Multiple Hypotheses in Behavioral Ecology." *Behavioral Ecology and Sociobiology* 65 (1): 37–45.

Dodds, Walter. 2009. *Laws, Theories, and Patterns in Ecology*. University of California Press.

Dorling, Danny. 2019. "Kindness: A New Kind of Rigour for British Geographers." *Emotion, Space and Society* 33 (November): 100630. https://doi.org/10.1016/j.emospa.2019.100630.

Dornelas, Maria, Nicholas J. Gotelli, Brian McGill, Hideyasu Shimadzu, Faye Moyes, Caya Sievers, and Anne E. Magurran. 2014. "Assemblage Time Series Reveal Biodiversity Change but Not Systematic Loss." *Science* 344 (6181): 296–99. https://doi.org/10.1126/science.1248484.

Douven, Igor. 2021. "Abduction." In *The Stanford Encyclopedia of Philosophy*, Summer 2021 ed., edited by Edward N. Zalta. Metaphysics Research Lab, Stanford University. https://plato.stanford.edu/archives/sum2021/entries/abduction/.

Downes, Barbara J. 2010. "Back to the Future: Little-Used Tools and Principles of Scientific Inference Can Help Disentangle Effects of Multiple Stressors on Freshwater Ecosystems." *Freshwater Biology* 55 (S1): S60–S79. https://doi.org/10.1111/j.1365-2427.2009.02377.x.

Drake, D. C., B. Maritz, S. M. Jacobs, C. J. Crous, A. Engelbrecht, A. Etale, M. J. Fourie, et al. 2013. "The Propagation and Dispersal of Misinformation in Ecology: Is There a Relationship between Citation Accuracy and Journal Impact Factor?" *Hydrobiologia* 702 (1): 1–4. https://doi.org/10.1007/s10750-012-1392-6.

Drake, John M., and Andrew M. Kramer. 2012. "Mechanistic Analogy: How Microcosms Explain Nature." *Theoretical Ecology* 5 (3): 433–44. https://doi.org/10.1007/s12080-011-0134-0.

Dresow, Max. 2023. "Ediacaran Enigma: Uncertainty and Underdetermination in Precambrian Paleontology." *Extinct* (blog), August 14. http://www.extinctblog.org/extinct/2023/8/14/ediacaran-enigma.

———. 2024a. "Release the Kraken." *Extinct* (blog), July 5. http://www.extinctblog.org/extinct/2024/7/5/release-the-kraken.

———. 2024b. "Wild Thing." *Extinct* (blog), July 12. http://www.extinctblog.org/extinct/2024/7/12/wild-thing.

Drotos, Katherine H. I., Douglas W. Larson, and R. Troy McMullin. 2024. "Scientific Telephone: The Cautionary Tale of the Global Coverage of Lichens." *BioScience* 74 (7): 473–77. https://doi.org/10.1093/biosci/biae048.

Duffy, Meghan. 2013. "Prioritizing Manuscripts, and Having Data Go Unpublished for Lack of Time." *Dynamic Ecology* (blog), April 16. https://dynamicecology.wordpress.com/2013/04/16/prioritizing-manuscripts-and-having-data-go-unpublished-for-lack-of-time/.

———. 2015. "Is the PEG Model Paper an Indicator of Changing Authorship Criteria?" *Dynamic Ecology* (blog), February 12. https://dynamicecology.wordpress.com/2015/02/12/is-the-peg-model-paper-an-indicator-of-changing-authorship-criteria/.

———. 2017. "Last and Corresponding Authorship Practices in Ecology." *Ecology and Evolution* 7 (21): 8876–87. https://doi.org/10.1002/ece3.3435.

Duffy, Meghan A., and Spencer R. Hall. 2008. "Selective Predation and Rapid Evolution Can Jointly Dampen Effects of Virulent Parasites on *Daphnia* Populations." *American Naturalist* 171 (4): 499–510. https://doi.org/10.1086/528998.

Dunne, Jennifer A., Richard J. Williams, and Neo D. Martinez. 2002. "Food-Web Structure and Network Theory: The Role of Connectance and Size." *Proceedings of the National Academy of Sciences* 99 (20): 12917–22. https://doi.org/10.1073/pnas.192407699.

Dyson, Freeman J. 2015. *Birds and Frogs: Selected Papers of Freeman Dyson, 1990–2014*. World Scientific Publishing.

Eadie, John McA., Louis Broekhoven, and Patrick Colgan. 1987. "Size Ratios and Artifacts: Hutchinson's Rule Revisited." *American Naturalist* 129 (1): 1–17.

Efron, Bradley. 2010. "The Future of Indirect Evidence." *Statistical Science: A Review Journal of the Institute of Mathematical Statistics* 25 (2): 145–57. https://doi.org/10.1214/09-STS308.

Efron, Bradley, and Carl Morris. 1977. "Stein's Paradox in Statistics." *Scientific American* 236 (5): 119–27.

Egerton, Frank N. 2013. "History of Ecological Sciences, Part 47: Ernst Haeckel's Ecology." *Bulletin of the Ecological Society of America* 94 (3): 222–44. https://doi.org/10.1890/0012-9623-94.3.222.

Egger, Matthias, George Davey Smith, Martin Schneider, and Christoph Minder. 1997. "Bias in Meta-analysis Detected by a Simple, Graphical Test." *BMJ* 315 (7109): 629–34. https://doi.org/10.1136/bmj.315.7109.629.

Eichhorn, Markus. 2016. "A Lie about My Childhood." *Trees in Space* (blog), April 6. https://treesinspace.com/2016/04/06/where-did-it-all-begin/.

Ellner, Stephen P. 2006. "How to Write a Theoretical Ecology Paper That People Will Cite." http://ellner.eeb.cornell.edu/CitedPapers.pdf.

Ellner, Stephen P., Robin E. Snyder, Peter B. Adler, and Giles Hooker. 2019. "An Expanded Modern Coexistence Theory for Empirical Applications." *Ecology Letters* 22 (1): 3–18. https://doi.org/10.1111/ele.13159.

Ellstrand, Norman C. 1983. "Why Are Juveniles Smaller Than Their Parents?" *Evolution* 37 (5): 1091–94. https://doi.org/10.2307/2408423.

Elton, Charles. 1935. "Eppur Si Muove." Edited by Alfred J. Lotka and S. A. Severtzoff. *Journal of Animal Ecology* 4 (1): 148–50. https://doi.org/10.2307/1225.

———. 1958. *The Ecology of Invasions by Animals and Plants*. Springer.

Engelke, Matthew. 2018. *How to Think Like an Anthropologist*. Princeton University Press.

Enserink, Martin. 2021. "Sea of Doubts." *Science* 372 (6542): 560–65. https://doi.org/10.1126/science.372.6542.560.

———. 2022. "Star Marine Ecologist Guilty of Misconduct, University Says." *Science* 377 (6607): 699–700. https://doi.org/10.1126/science.ade3374.

Evans, Matthew R., Volker Grimm, Karin Johst, Tarja Knuuttila, Rogier de Langhe, Catherine M. Lessells, Martina Merz, et al. 2013. "Do Simple Models Lead to Generality in Ecology?" *Trends in Ecology & Evolution* 28 (10): 578–83. https://doi.org/10.1016/j.tree.2013.05.022.

Ezenwa, Vanessa O., Rampal S. Etienne, Gordon Luikart, Albano Beja-Pereira, and Anna E. Jolles. 2010. "Hidden Consequences of Living in a Wormy World: Nematode-Induced Immune Suppression Facilitates Tuberculosis Invasion in African Buffalo." *American Naturalist* 176 (5): 613–24. https://doi.org/10.1086/656496.

Ezenwa, Vanessa O., and Anna E. Jolles. 2015. "Opposite Effects of Anthelmintic Treatment on Microbial Infection at Individual versus Population Scales." *Science* 347 (6218): 175–77. https://doi.org/10.1126/science.1261714.

Fahrig, Lenore. 2017a. "Ecological Responses to Habitat Fragmentation Per Se." *Annual Review of Ecology, Evolution, and Systematics* 48 (2017): 1–23. https://doi.org/10.1146/annurev-ecolsys-110316-022612.

———. 2017b. "Forty Years of Bias in Habitat Fragmentation Research." In *Effective Conservation Science: Data Not Dogma*, edited by Peter Kareiva, Michelle Marvier, and Brian Silliman, 32–38. Oxford University Press. https://doi.org/10.1093/oso/9780198808978.003.0005.

Fahrig, Lenore, James I. Watling, Carlos Alberto Arnillas, Víctor Arroyo-Rodríguez, Theresa Jörger-Hickfang, Jörg Müller, Henrique M. Pereira, et al. 2022. "Resolving the SLOSS Dilemma for Biodiversity Conservation: A Research Agenda." *Biological Reviews of the Cambridge Philosophical Society* 97 (1): 99–114. https://doi.org/10.1111/brv.12792.

Fauth, J. E., J. Bernardo, M. Camara, Resetarits W. J., J. Van Buskirk, and S. A. McCollum. 1996. "Simplifying the Jargon of Community Ecology: A Conceptual Approach." *American Naturalist* 147 (2): 282–86. https://doi.org/10.1086/285850.

Fawcett, Tim W., and Andrew D. Higginson. 2012. "Heavy Use of Equations Impedes Communication among Biologists." *Proceedings of the National Academy of Sciences* 109 (29): 11735–39. https://doi.org/10.1073/pnas.1205259109.

Ferguson, Neil M., Christl A. Donnelly, and Roy M. Anderson. 2001. "The Foot-and-Mouth Epidemic in Great Britain: Pattern of Spread and Impact of Interventions." *Science* 292 (5519): 1155–60. https://doi.org/10.1126/science.1061020.

Fernández-Llamazares, Álvaro, Julia E. Fa, Dan Brockington, Eduardo S. Brondízio, Joji Cariño, Esteve Corbera, Maurizio Farhan Ferrari, et al. 2024. "No Basis for Claim That 80% of Biodiversity Is Found in Indigenous Territories." *Nature* 633 (8028): 32–35. https://doi.org/10.1038/d41586-024-02811-w.

Foley, Carolyn J., Zachary S. Feiner, Timothy D. Malinich, and Tomas O. Höök. 2018. "A Meta-analysis of the Effects of Exposure to Microplastics on Fish and Aquatic Invertebrates." *Science of the Total Environment* 631–32:550–59. https://doi.org/10.1016/j.scitotenv.2018.03.046.

Ford, Adam T., and Jacob R. Goheen. 2015. "Trophic Cascades by Large Carnivores: A Case for Strong Inference and Mechanism." *Trends in Ecology & Evolution* 30 (12): 725–35. https://doi.org/10.1016/j.tree.2015.09.012.

Fordham, Damien A., Cleo Bertelsmeier, Barry W. Brook, Regan Early, Dora Neto, Stuart C. Brown, Sébastien Ollier, and Miguel B. Araújo. 2018. "How Complex Should Models Be? Comparing Correlative and Mechanistic Range Dynamics Models." *Global Change Biology* 24 (3): 1357–70. https://doi.org/10.1111/gcb.13935.

Fourcade, Yoan, Aurélien G. Besnard, and Jean Secondi. 2018. "Paintings Predict the Distribution of Species, or the Challenge of Selecting Environmental Predictors and Evaluation Statistics." *Global Ecology and Biogeography* 27 (2): 245–56. https://doi.org/10.1111/geb.12684.

Fournier, Auriel. 2016. "How I Got Here." May 24. https://aurielfournier.github.io/how-i-got-here/.

Fox, Charles W. 2017. "Difficulty of Recruiting Reviewers Predicts Review Scores and Editorial Decisions at Six Journals of Ecology and Evolution." *Scientometrics* 113 (1): 465–77. https://doi.org/10.1007/s11192-017-2489-5.

———. 2021. "Which Peer Reviewers Voluntarily Reveal Their Identity to Authors? Insights into the Consequences of Open-Identities Peer Review." *Proceedings of the Royal Society B: Biological Sciences* 288 (1961): 20211399. https://doi.org/10.1098/rspb.2021.1399.

Fox, Charles W., C. Sean Burns, Anna D. Muncy, and Jennifer A. Meyer. 2017. "Author-Suggested Reviewers: Gender Differences and Influences on the Peer Review Process at an Ecology Journal." *Functional Ecology* 31 (1): 270–80. https://doi.org/10.1111/1365-2435.12665.

Fox, Charles W., Meghan A. Duffy, Daphne J. Fairbairn, and Jennifer A. Meyer. 2019. "Gender Diversity of Editorial Boards and Gender Differences in the Peer Review Process at Six Journals of Ecology and Evolution." *Ecology and Evolution* 9 (24): 13636–49. https://doi.org/10.1002/ece3.5794.

Fox, Charles W., Jennifer Meyer, and Emilie Aimé. 2023. "Double-Blind Peer Review Affects Reviewer Ratings and Editor Decisions at an Ecology Journal." *Functional Ecology* 37 (5): 1144–57. https://doi.org/10.1111/1365-2435.14259.

Fox, Charles W., and C. E. Timothy Paine. 2019. "Gender Differences in Peer Review Outcomes and Manuscript Impact at Six Journals of Ecology and Evolution." *Ecology and Evolution* 9 (6): 3599–3619. https://doi.org/10.1002/ece3.4993.

Fox, Charles W., C. E. Timothy Paine, and Boris Sauterey. 2016. "Citations Increase with Manuscript Length, Author Number, and References Cited in Ecology Journals." *Ecology and Evolution* 6 (21): 7717–26. https://doi.org/10.1002/ece3.2505.

Fox, Jeremy W. 2002. "Testing a Simple Rule for Dominance in Resource Competition." *American Naturalist* 159 (3): 305–19. https://doi.org/10.1086/338543.

———. 2005. "Interpreting the 'Selection Effect' of Biodiversity on Ecosystem Function." *Ecology Letters* 8 (8): 846–56. https://doi.org/10.1111/j.1461-0248.2005.00795.x.

———. 2006. "Using the Price Equation to Partition the Effects of Biodiversity Loss on Ecosystem Function." *Ecology* 87 (11): 2687–96. https://doi.org/10.1890/0012-9658(2006)87[2687:UTPETP]2.0.CO;2.

———. 2007. "The Dynamics of Top-Down and Bottom-Up Effects in Food Webs of Varying Prey Diversity, Composition, and Productivity." *Oikos* 116 (2): 189–200. https://doi.org/10.1111/j.0030-1299.2007.15280.x.

———. 2011a. "Advice: How I Almost Quit Science." *Dynamic Ecology* (blog), July 19. https://dynamicecology.wordpress.com/2011/07/19/advice-how-i-almost-quit-science/.

———. 2011b. "Objections to Microcosms in Ecology, and Their Answers." *Dynamic Ecology* (blog), June 10. https://dynamicecology.wordpress.com/2011/06/10/objections-to-microcosms-in-ecology-and-their-answers/.

———. 2012a. "Advice: 20 Different Stability Concepts | Dynamic Ecology." *Dynamic Ecology* (blog), May 16. https://dynamicecology.wordpress.com/2012/05/16/advice-20-different-stability-concepts/.

———. 2012b. "Can Blogging Change How Ecologists Share Ideas? In Economics, It Already Has." *Ideas in Ecology and Evolution* 5 (2). https://ojs.library.queensu.ca/index.php/IEE/article/view/4457.

———. 2012c. "The Fox and the Hedgehog." *Dynamic Ecology* (blog), July 9. https://dynamicecology.wordpress.com/2012/07/09/the-fox-and-the-hedgehog/.

———. 2013a. "E. O. Wilson vs. Math (UPDATEDx16 & Corrected; No Further Updates)." *Dynamic Ecology* (blog), April 7. https://dynamicecology.wordpress.com/2013/04/07/e-o-wilson-vs-math/.

———. 2013b. "The Intermediate Disturbance Hypothesis Should Be Abandoned." *Trends in Ecology & Evolution* 28 (2): 86–92. https://doi.org/10.1016/j.tree.2012.08.014.

———. 2013c. "What Metacommunity Ecology Can Learn from Population Genetics." *Dynamic Ecology* (blog), 2013. https://dynamicecology.wordpress.com/2013/10/17/what-metacommunity-ecology-can-learn-from-population-genetics/.

———. 2016a. "How I Got into Ecology." *Dynamic Ecology* (blog), May 24. https://dynamicecology.wordpress.com/2016/05/24/how-i-got-into-ecology/.

———. 2016b. "The Microcosm Wars Are Long Since Over. The Microcosmologists Won." *Dynamic Ecology* (blog), June 20. https://dynamicecology.wordpress.com/2016/06/20/the-microcosm-wars-are-long-since-over-the-microcosmologists-won/.

———. 2017. "Is Ecology a Single Coherent Scientific Discipline? (Includes Poll)." *Dynamic Ecology* (blog), September 26. https://dynamicecology.wordpress.com/2017/09/26/is-ecology-a-single-scientific-discipline/.

———. 2018. "Poll Results: Here's What Our Readers Think about Some of the Most Controversial Ideas in Ecology." *Dynamic Ecology* (blog), April 30. https://dynamicecology.wordpress.com/2018/04/30/poll-results-heres-what-our-readers-think-about-some-of-the-most-controversial-ideas-in-ecology/.

———. 2019a. "Poll Results: The Many Ways Ecologists Seek Generality (and Why Some Are Much More Popular Than Others)." *Dynamic Ecology* (blog), November 4. https://dynamicecology.wordpress.com/2019/11/04/poll-results-the-many-ways-ecologists-seek-generality/.

———. 2019b. "Poll Results: What Are the Biggest Problems with the Conduct of Ecological Research?" *Dynamic Ecology* (blog), June 3. https://dynamicecology.wordpress.com/2019/06/03/poll-results-what-are-the-biggest-problems-with-the-conduct-of-ecological-research/.

———. 2020a. "2020 Reader Survey Results." *Dynamic Ecology* (blog), January 29. https://dynamicecology.wordpress.com/2020/01/29/2020-reader-survey-results/.

———. 2020b. "Dynamic Ecology Year in Review." *Dynamic Ecology* (blog), January 2. https://dynamicecology.wordpress.com/2020/01/02/dynamic-ecology-year-in-review-4/.

———. 2022. "How Much Does the Typical Ecological Meta-analysis Overestimate the True Mean Effect Size?" *Ecology and Evolution* 12 (11): e9521. https://doi.org/10.1002/ece3.9521.

———. 2023. "The Existence and Strength of Higher Order Interactions Is Sensitive to Environmental Context." *Ecology* 104 (10): e4156. https://doi.org/10.1002/ecy.4156.

Fox, Jeremy W., and Benjamin Kerr. 2012. "Analyzing the Effects of Species Gain and Loss on Ecosystem Function Using the Extended Price Equation Partition." *Oikos* 121 (2): 290–98.

Fox, Jeremy W., and Richard E. Lenski. 2015. "From Here to Eternity—the Theory and Practice of a Really Long Experiment." *PLOS Biology* 13 (6): e1002185. https://doi.org/10.1371/journal.pbio.1002185.

Fox, Jeremy W., and Diane Srivastava. 2006. "Predicting Local–Regional Richness Relationships Using Island Biogeography Models." *Oikos* 113 (2): 376–82. https://doi.org/10.1111/j.2006.0030-1299.14768.x.

Frank, Steven A. 2014. "Generative Models versus Underlying Symmetries to Explain Biological Pattern." *Journal of Evolutionary Biology* 27 (6): 1172–78. https://doi.org/10.1111/Jeb.12388.

Frank, Steven A., and Jordi Bascompte. 2019. "Invariance in Ecological Pattern." *F1000Research* 8. https://www.ncbi.nlm.nih.gov/pmc/articles/PMC7008606/.

Frank, Steven A., and William Godsoe. 2020. "The Generalized Price Equation: Forces That Change Population Statistics." *Frontiers in Ecology and Evolution* 8 (August). https://doi.org/10.3389/fevo.2020.00240.

Frankel, Henry R. 2012. *The Continental Drift Controversy: Wegener and the Early Debate*. Cambridge University Press.

Fraser, Hannah, Ashley Barnett, Timothy H. Parker, and Fiona Fidler. 2020. "The Role of Replication Studies in Ecology." *Ecology and Evolution* 10 (12): 5197–5207. https://doi.org/10.1002/ece3.6330.

Fraser, Hannah, Tim Parker, Shinichi Nakagawa, Ashley Barnett, and Fiona Fidler. 2018. "Questionable Research Practices in Ecology and Evolution." *PLOS ONE* 13 (7): e0200303. https://doi.org/10.1371/journal.pone.0200303.

Fraser, Lauchlan H., Hugh A. L. Henry, Cameron N. Carlyle, Shannon R. White, Carl Beierkuhnlein, James F. Cahill Jr., Brenda B. Casper, et al. 2013. "Coordinated Distributed Experiments: An Emerging Tool for Testing Global

Hypotheses in Ecology and Environmental Science." *Frontiers in Ecology and the Environment* 11 (3): 147–55. https://doi.org/10.1890/110279.

Freckleton, Robert P., Andrew R. Watkinson, and Mark Rees. 2009. "Measuring the Importance of Competition in Plant Communities." *Journal of Ecology* 97 (3): 379–84. https://doi.org/10.1111/j.1365-2745.2009.01497.x.

Freilich, Mara A., Evie Wieters, Bernardo R. Broitman, Pablo A. Marquet, and Sergio A. Navarrete. 2018. "Species Co-occurrence Networks: Can They Reveal Trophic and Non-trophic Interactions in Ecological Communities?" *Ecology* 99 (3): 690–99. https://doi.org/10.1002/ecy.2142.

Fretwell, Stephen D. 1975. "The Impact of Robert Macarthur on Ecology." *Annual Review of Ecology and Systematics* 6 (1): 1–13. https://doi.org/10.1146/annurev.es.06.110175.000245.

Fuentes, Agustin. 2016. "The Extended Evolutionary Synthesis, Ethnography, and the Human Niche: Toward an Integrated Anthropology." *Current Anthropology* 57 (S13): S13–S26. https://doi.org/10.1086/685684.

Fusco, Emily J., Brian J. Falk, Paul J. Heimowitz, Deah Lieurance, Elliot W. Parsons, Cait M. Rottler, Lindsey L. Thurman, and Annette G. Evans. 2024. "The Emerging Invasive Species and Climate Change Lexicon." *Trends in Ecology & Evolution* 39:1119–29. https://doi.org/10.1016/j.tree.2024.08.005.

Gause, G. F. 2019. *The Struggle for Existence*. Courier Dover Publications.

"GDP per Capita." n.d. Our World in Data. Accessed October 24, 2024. https://ourworldindata.org/grapher/gdp-per-capita-worldbank.

Gelman, Andrew, and John Carlin. 2014. "Beyond Power Calculations: Assessing Type S (Sign) and Type M (Magnitude) Errors." *Perspectives on Psychological Science: A Journal of the Association for Psychological Science* 9 (6): 641–51. https://doi.org/10.1177/1745691614551642.

Gibbs, James Lowell Jr. 1998. "Stanford Anthropology Department Splits." *Anthropology News* 39 (7): 21. https://doi.org/10.1111/an.1998.39.7.21.2.

Gilbert, Benjamin, and Joseph R. Bennett. 2010. "Partitioning Variation in Ecological Communities: Do the Numbers Add Up?" *Journal of Applied Ecology* 47 (5): 1071–82. https://doi.org/10.1111/j.1365-2664.2010.01861.x.

Gingerich, Philip D. 2006. "Environment and Evolution through the Paleocene–Eocene Thermal Maximum." *Trends in Ecology & Evolution* 21 (5): 246–53. https://doi.org/10.1016/j.tree.2006.03.006.

Glazier, Douglas S. 2014. "Metabolic Scaling in Complex Living Systems." *Systems* 2 (4): 451–540. https://doi.org/10.3390/systems2040451.

Gonçalves-Souza, Thiago, Gustavo Q. Romero, and Karl Cottenie. 2013. "A Critical Analysis of the Ubiquity of Linear Local–Regional Richness Relationships." *Oikos* 122 (7): 961–66. https://doi.org/10.1111/j.1600-0706.2013.00305.x.

Gonzalez, Andrew, Bradley J. Cardinale, Ginger R. H. Allington, Jarrett Byrnes, K. Arthur Endsley, Daniel G. Brown, David U. Hooper, Forest Isbell, Mary I. O'Connor, and Michel Loreau. 2016. "Estimating Local Biodiversity Change: A Critique of Papers Claiming No Net Loss of Local Diversity." *Ecology* 97 (8): 1949–60. https://doi.org/10.1890/15-1759.1.

Gonzalez, Andrew, Rachel M. Germain, Diane S. Srivastava, Elise Filotas, Laura E. Dee, Dominique Gravel, Patrick L. Thompson, et al. 2020. "Scaling-Up Biodiversity-Ecosystem Functioning Research." *Ecology Letters* 23 (4): 757–76. https://doi.org/10.1111/ele.13456.

Goodman, Daniel. 1975. "The Theory of Diversity-Stability Relationships in Ecology." *Quarterly Review of Biology* 50 (3): 237–66. https://doi.org/10.1086/408563.

Gorham, Eville, and Julia Kelly. 2014. "Multiauthorship, an Indicator of the Trend toward Team Research in Ecology." *Bulletin of the Ecological Society of America* 95 (3): 243–49. https://doi.org/10.1890/0012-9623-95.3.243.

Gossett, Thomas F. 1997. *Race: The History of an Idea in America*. Oxford University Press USA.

Gotelli, Nicholas J., and Gary R. Graves. 1996. *Null Models in Ecology*. Smithsonian Institution Press. https://repository.si.edu/xmlui/handle/10088/7782.

Gould, Elliot, Hannah S. Fraser, Timothy H. Parker, Shinichi Nakagawa, Simon C. Griffith, Peter A. Vesk, Fiona Fidler, et al. 2025. "Same Data, Different Analysts: Variation in Effect Sizes Due to Analytical Decisions in Ecology and Evolutionary Biology." *BMC Biology* 23: article 35. https://doi.org/10.1186/s12915-024-02101-x.

Gould, Stephen Jay. 1980. "Is a New and General Theory of Evolution Emerging?" *Paleobiology* 6 (1): 119–30.

Gould, Stephen J., and R. C. Lewontin. 1979. "The Spandrels of San Marco and the Panglossian Paradigm: A Critique of the Adaptationist Programme." *Proceedings of the Royal Society B: Biological Sciences* 205 (1161): 581–98. https://doi.org/10.1098/rspb.1979.0086.

Govaert, Lynn, Jelena H. Pantel, and Luc De Meester. 2016. "Eco-evolutionary Partitioning Metrics: Assessing the Importance of Ecological and Evolutionary Contributions to Population and Community Change." *Ecology Letters* 19 (8): 839–53. https://doi.org/10.1111/ele.12632.

Gowers, W. Timothy. 2000. "The Two Cultures of Mathematics." https://www.dpmms.cam.ac.uk/~wtg10/2cultures.pdf.

Grace, James B. 2024. "An Integrative Paradigm for Building Causal Knowledge." *Ecological Monographs* 94 (4): e1628. https://doi.org/10.1002/ecm.1628.

Grainger, Tess N., Athmanathan Senthilnathan, Po-Ju Ke, Matthew A. Barbour, Natalie T. Jones, John P. DeLong, Sarah P. Otto, et al. 2022. "An Empiricist's

Guide to Using Ecological Theory." *American Naturalist* 199 (1): 1–20. https://doi.org/10.1086/717206.

Grant, Peter R., and B. Rosemary Grant. 2014. *40 Years of Evolution: Darwin's Finches on Daphne Major Island*. Illus. ed. Princeton University Press.

Gremer, Jennifer R., and D. Lawrence Venable. 2014. "Bet Hedging in Desert Winter Annual Plants: Optimal Germination Strategies in a Variable Environment." *Ecology Letters* 17 (3): 380–87. https://doi.org/10.1111/ele.12241.

Grilli, Jacopo, Matteo Adorisio, Samir Suweis, György Barabás, Jayanth R. Banavar, Stefano Allesina, and Amos Maritan. 2017. "Feasibility and Coexistence of Large Ecological Communities." *Nature Communications* 8 (1): 14389. https://doi.org/10.1038/ncomms14389.

Grimm, Volker. 1994. "Mathematical Models and Understanding in Ecology." In "State-of-the-Art in Ecological Modelling Proceedings of ISEM's 8th International Conference," special issue, *Ecological Modelling* 75–76 (September): 641–51. https://doi.org/10.1016/0304-3800(94)90056-6.

Grimm, Volker, and Christian Wissel. 1997. "Babel, or the Ecological Stability Discussions: An Inventory and Analysis of Terminology and a Guide for Avoiding Confusion." *Oecologia* 109 (3): 323–34.

Grogan, Paul. 2005. "The Use of Hypotheses in Ecology." *Bulletin of the British Ecological Society* 36:43–47.

Gross, Rachael B., and Robin Heinsohn. 2023. "Elephants Not in the Room: Systematic Review Shows Major Geographic Publication Bias in African Elephant Ecological Research." *Diversity* 15:451. https://doi.org/10.3390/d15030451.

Guerin, Greg R., Irene Martín-Forés, Ben Sparrow, and Andrew J. Lowe. 2018. "The Biodiversity Impacts of Non-native Species Should Not Be Extrapolated from Biased Single-Species Studies." *Biodiversity and Conservation* 27 (3): 785–90. https://doi.org/10.1007/s10531-017-1439-0.

Gurevitch, Jessica, Laura L. Morrow, Alison Wallace, and Joseph S. Walsh. 1992. "A Meta-analysis of Competition in Field Experiments." *American Naturalist* 140 (4): 539–72.

Haack, Susan. 2000. *Manifesto of a Passionate Moderate: Unfashionable Essays*. University of Chicago Press.

Haigh, Martin J., and T. W. Freeman. 1982. "The Crisis in American Geography." *Area* 14 (3): 185–90.

Haller, Benjamin C. 2014. "Theoretical and Empirical Perspectives in Ecology and Evolution: A Survey." *BioScience* 64 (10): 907–16. https://doi.org/10.1093/biosci/biu131.

Hammal, Omar Al, David Alonso, Rampal S. Etienne, and Stephen J. Cornell. 2015. "When Can Species Abundance Data Reveal Non-neutrality?" *PLOS Computational Biology* 11 (3): e1004134. https://doi.org/10.1371/journal.pcbi.1004134.

Hanski, Ilkka. 1982. "Dynamics of Regional Distribution: The Core and Satellite Species Hypothesis." *Oikos* 38 (2): 210–21. https://doi.org/10.2307/3544021.

Hargreaves, Anna L., Karen E. Samis, and Christopher G. Eckert. 2014. "Are Species' Range Limits Simply Niche Limits Writ Large? A Review of Transplant Experiments beyond the Range." *American Naturalist* 183 (2): 157–73. https://doi.org/10.1086/674525.

Harrer, Mathias, Pim Cuijpers, Toshi A. Furukawa, and David D. Ebert. 2021. *Doing Meta-analysis with R: A Hands-On Guide*. Chapman & Hall.

Harrison, Gary W. 1995. "Comparing Predator-Prey Models to Luckinbill's Experiment with Didinium and Paramecium." *Ecology* 76 (2): 357–74. https://doi.org/10.2307/1941195.

Harrison, Natasha Dean, and Ella L. Kelly. 2022. "Affordable RFID Loggers for Monitoring Animal Movement, Activity, and Behaviour." *PLOS ONE* 17 (10): e0276388. https://doi.org/10.1371/journal.pone.0276388.

Harte, John. 2014. "Diverse Introspectives: A Conversation with John Harte." Interview by Jes Coyle. BioDiverse Perspectives, August 31. Available at http://web.archive.org/web/20140922113514/http://www.biodiverseperspectives.com/2014/08/31/diverse-introspectives-a-conversation-with-john-harte/.

Hastings, Alan. 1987. "Can Competition Be Detected Using Species Co-occurrence Data?" *Ecology* 68 (1): 117–23. https://doi.org/10.2307/1938811.

Hawkins, Bradford A., José Alexandre Felizola Diniz-Filho, and Arthur E. Weis. 2005. "The Mid-domain Effect and Diversity Gradients: Is There Anything to Learn?" *American Naturalist* 166 (5): E140–43. https://doi.org/10.1086/491686.

Haydon, Daniel T., Rowland R. Kao, and R. Paul Kitching. 2004. "The UK Foot-and-Mouth Disease Outbreak—the Aftermath." *Nature Reviews Microbiology* 2 (8): 675–81. https://doi.org/10.1038/nrmicro960.

He, Fangliang, and Stephen P. Hubbell. 2011. "Species–Area Relationships Always Overestimate Extinction Rates from Habitat Loss." *Nature* 473 (7347): 368–71. https://doi.org/10.1038/nature09985.

Healy, Kieran. 2018. *Data Visualization: A Practical Introduction*. Princeton University Press.

Hector, Andy, and Robert Bagchi. 2007. "Biodiversity and Ecosystem Multifunctionality." *Nature* 448 (7150): 188–90. https://doi.org/10.1038/nature05947.

Hector, Andy, and Rowan Hooper. 2002. "Darwin and the First Ecological Experiment." *Science* 295 (5555): 639–40. https://doi.org/10.1126/science.1064815.

Hedges, Larry V., Jessica Gurevitch, and Peter S. Curtis. 1999. "The Meta-analysis of Response Ratios in Experimental Ecology." *Ecology* 80 (4):

1150–56. https://doi.org/10.1890/0012-9658(1999)080[1150:TMAORR]2.0.CO;2.

Hendry, Andrew. 2014. "How to Write/Present Science: BABY-WEREWOLF-SILVER BULLET." *Eco-evo Evo-eco* (blog), October 18. http://ecoevoevoeco.blogspot.com/2014/10/how-to-writepresent-science-baby.html.

Hengeveld, R. 1992. "Special Comment: Right and Wrong in Ecological Explanation." *Journal of Biogeography* 19 (4): 345–47. https://doi.org/10.2307/2845561.

Higgins, Julian P. T., and Simon G. Thompson. 2002. "Quantifying Heterogeneity in a Meta-Analysis." *Statistics in Medicine* 21 (11): 1539–58. https://doi.org/10.1002/sim.1186.

Hillebrand, Helmut. 2005. "Regressions of Local on Regional Diversity Do Not Reflect the Importance of Local Interactions or Saturation of Local Diversity." *Oikos* 110 (1): 195–98. https://doi.org/10.1111/j.0030-1299.2005.14008.x.

Hintzen, Rogier E., Marina Papadopoulou, Ross Mounce, Cristina Banks-Leite, Robert D. Holt, Morena Mills, Andrew T Knight, Armand M. Leroi, and James Rosindell. 2020. "Relationship between Conservation Biology and Ecology Shown through Machine Reading of 32,000 Articles." *Conservation Biology* 34 (3): 721–32. https://doi.org/10.1111/cobi.13435.

Hodges, Karen E. 2008. "Defining the Problem: Terminology and Progress in Ecology." *Bulletin of the Ecological Society of America* 6:35–42. https://doi.org/10.1890/060108.

Hoeksema, Jason D., V. Bala Chaudhary, Catherine A. Gehring, Nancy Collins Johnson, Justine Karst, Roger T. Koide, Anne Pringle, et al. 2010. "A Meta-analysis of Context-Dependency in Plant Response to Inoculation with Mycorrhizal Fungi." *Ecology Letters* 13 (3): 394–407. https://doi.org/10.1111/j.1461-0248.2009.01430.x.

Holling, C. S. 1966. "The Strategy of Building Models of Complex Ecological Systems." In *Systems Analysis in Ecology*, edited by Kenneth E. S. Watt, 195–214. Academic Press.

Hong, Pubin, Bernhard Schmid, Frederik De Laender, Nico Eisenhauer, Xingwen Zhang, Haozhen Chen, Dylan Craven, et al. 2022. "Biodiversity Promotes Ecosystem Functioning despite Environmental Change." *Ecology Letters* 25 (2): 555–69. https://doi.org/10.1111/ele.13936.

Horn, Henry S., and Robert M. May. 1977. "Limits to Similarity among Coexisting Competitors." *Nature* 270 (5639): 660–61. https://doi.org/10.1038/270660a0.

Hossenfelder, Sabine. 2018. *Lost in Math: How Beauty Leads Physics Astray*. Illus. ed. Basic Books.

Hubbell, Stephen P. 2001. *The Unified Neutral Theory of Biodiversity and Biogeography (MPB-32)*. Princeton University Press.

Huber, Juergen, Sabiou Inoua, Rudolf Kerschbamer, Christian König-Kersting, Stefan Palan, and Vernon L. Smith. 2022. "Nobel and Novice: Author Prominence Affects Peer Review." SSRN Scholarly Paper. Rochester, New York. https://doi.org/10.2139/ssrn.4190976.

Hull, David L. 1990. *Science as a Process: An Evolutionary Account of the Social and Conceptual Development of Science*. Science and Its Conceptual Foundations Series. University of Chicago Press.

Hulme, Philip E. 2015. "Invasion Pathways at a Crossroad: Policy and Research Challenges for Managing Alien Species Introductions." *Journal of Applied Ecology* 52 (6): 1418–24. https://doi.org/10.1111/1365-2664.12470.

Hulme, Philip E., Petr Pyšek, Vojtěch Jarošík, Jan Pergl, Urs Schaffner, and Montserrat Vilà. 2013. "Bias and Error in Understanding Plant Invasion Impacts." *Trends in Ecology & Evolution* 28 (4): 212–18. https://doi.org/10.1016/j.tree.2012.10.010.

Huntington-Klein, Nick, Andreu Arenas, Emily Beam, Marco Bertoni, Jeffrey R. Bloem, Pralhad Burli, Naibin Chen, et al. 2021. "The Influence of Hidden Researcher Decisions in Applied Microeconomics." *Economic Inquiry* 59 (3): 944–60. https://doi.org/10.1111/ecin.12992.

Hurst, Michael E. Eliot. 1985. "Geography Has Neither Existence nor Future." In *The Future of Geography*, edited by Ron Johnson, 59–91. Routledge Library Editions: Social and Cultural Geography. Routledge.

Huston, Michael A. 1997. "Hidden Treatments in Ecological Experiments: Re-evaluating the Ecosystem Function of Biodiversity." *Oecologia* 110 (4): 449–60. https://doi.org/10.1007/s004420050180.

Hutchinson, G. Evelyn. 1961. "The Paradox of the Plankton." *American Naturalist* 95 (882): 137–45.

———. 1978. *An Introduction to Population Ecology*. Yale University Press.

Hutchinson, G. Evelyn, and Robert H. MacArthur. 1959. "A Theoretical Ecological Model of Size Distributions among Species of Animals." *American Naturalist* 93 (869): 117–25. https://doi.org/10.1086/282063.

Huxley, Julian. 2010. *Evolution: The Modern Synthesis*. MIT Press.

Ishida, Yoichi. 2007. "Patterns, Models, and Predictions: Robert MacArthur's Approach to Ecology." *Philosophy of Science* 74 (5): 642–53. https://doi.org/10.1086/525610.

Ives, Anthony R. 2014. "Diverse Introspectives with Tony Ives." *Fletcher Halliday* (blog), March 11. https://fletcherhalliday.com/2014/03/11/diverse-introspectives-with-tony-ives/.

Ives, Anthony R., Árni Einarsson, Vincent A. A. Jansen, and Arnthor Gardarsson. 2008. "High-Amplitude Fluctuations and Alternative Dynamical States of Midges in Lake Myvatn." *Nature* 452 (7183): 84–87. https://doi.org/10.1038/nature06610.

Ives, Anthony R., Johannes Foufopoulos, Eric D. Klopfer, Jennifer L. Klug, and Todd M. Palmer. 1996. "Bottle or Big-Scale Studies: How Do We Do Ecology?" *Ecology* 77 (3): 681–85. https://doi.org/10.2307/2265491.

Ives, Anthony R., J. L. Klug, and K. Gross. 2000. "Stability and Species Richness in Complex Communities." *Ecology Letters* 3 (5): 399–411. https://doi.org/10.1046/j.1461-0248.2000.00144.x.

Jaschik, Scott. 2007. "A Forced Anthropology Merger." *Inside Higher Ed*, February 5. https://www.insidehighered.com/news/2007/02/06/forced-anthropology-merger.

Jennions, Michael D., and Anders Pape Møller. 2003. "A Survey of the Statistical Power of Research in Behavioral Ecology and Animal Behavior." *Behavioral Ecology* 14 (3): 438–45. https://doi.org/10.1093/beheco/14.3.438.

Jeschke, Jonathan, Lorena Gómez Aparicio, Sylvia Haider, Tina Heger, Christopher Lortie, Petr Pyšek, and David Strayer. 2012. "Support for Major Hypotheses in Invasion Biology Is Uneven and Declining." *NeoBiota* 14 (August): 1–20. https://doi.org/10.3897/neobiota.14.3435.

Jessup, Christine M., Samantha E. Forde, and Brendan J. M. Bohannan. 2005. "Microbial Experimental Systems in Ecology." In *Advances in Ecological Research*, vol. 37, *Population Dynamics and Laboratory Ecology*, edited by Robert A. Desharnais, 273–307. Academic Press. https://doi.org/10.1016/S0065-2504(04)37009-1.

Ji, Zhengyu, Yin Huang, Yao Feng, Anders Johansen, Jianming Xue, Louis A. Tremblay, and Zhaojun Li. 2021. "Effects of Pristine Microplastics and Nanoplastics on Soil Invertebrates: A Systematic Review and Meta-analysis of Available Data." *Science of the Total Environment* 788:147784. https://doi.org/10.1016/j.scitotenv.2021.147784.

Jiang, Lin. 2007. "Negative Selection Effects Suppress Relationships between Bacterial Diversity and Ecosystem Functioning." *Ecology* 88 (5): 1075–85. https://doi.org/10.1890/06-1556.

Jiang, Lin, Zhichao Pu, and Diana R. Nemergut. 2008. "On the Importance of the Negative Selection Effect for the Relationship between Biodiversity and Ecosystem Functioning." *Oikos* 117 (4): 488–93.

Johnson, Michele A., Ambika Kamath, Rebecca Kirby, Carla C. Fresquez, Su Wang, Chelsea M. Stehle, Alan R. Templeton, and Jonathan B. Losos. 2021. "What Determines Paternity in Wild Lizards? A Spatiotemporal Analysis of Behavior and Morphology." *Integrative and Comparative Biology* 61 (2): 634–42.

Johnston, Francis E., and Setha M. Low. 1984. "Biomedical Anthropology: An Emerging Synthesis in Anthropology." *American Journal of Physical Anthropology* 27 (S5): 215–27. https://doi.org/10.1002/ajpa.1330270510.

Jolles, Anna E., Vanessa O. Ezenwa, Rampal S. Etienne, Wendy C. Turner, and Han Olff. 2008. "Interactions between Macroparasites and Microparasites

Drive Infection Patterns in Free-Ranging African Buffalo." *Ecology* 89 (8): 2239–50. https://doi.org/10.1890/07-0995.1.

Jost, Lou. 2006. "Entropy and Diversity." *Oikos* 113 (2): 363–75. https://doi.org/10.1111/j.2006.0030-1299.14714.x.

Jurasinski, Gerald, Vroni Retzer, and Carl Beierkuhnlein. 2009. "Inventory, Differentiation, and Proportional Diversity: A Consistent Terminology for Quantifying Species Diversity." *Oecologia* 159 (1): 15–26. https://doi.org/10.1007/s00442-008-1190-z.

Justus, James. 2012. "The Elusive Basis of Inferential Robustness." *Philosophy of Science* 79 (5): 795–807. https://doi.org/10.1086/667902.

Kamath, Ambika, and Jonathan Losos. 2017. "The Erratic and Contingent Progression of Research on Territoriality: A Case Study." *Behavioral Ecology and Sociobiology* 71 (6): 89. https://doi.org/10.1007/s00265-017-2319-z.

———. 2018. "Reconsidering Territoriality Is Necessary for Understanding *Anolis* Mating Systems." *Behavioral Ecology and Sociobiology* 72 (7): 106. https://doi.org/10.1007/s00265-018-2524-4.

Kamath, Ambika, Beans Velocci, Ashton Wesner, Nancy Chen, Vince Formica, Banu Subramaniam, and María Rebolleda-Gómez. 2022. "Nature, Data, and Power: How Hegemonies Shaped This Special Section." *American Naturalist* 200 (1): 81–88. https://doi.org/10.1086/720001.

Kareiva, Peter. 1997. "EcoEssay Series Number 1—Response." National Center for Ecological Analysis and Synthesis. https://www.nceas.ucsb.edu/ecoessays/brown/kareiva.

———. 2014. "Renewing the Dialogue between Theory and Experiments in Population Ecology." In *Perspectives in Ecological Theory*, edited by Jonathan Roughgarden, Robert M. May, and Simon A. Levin, 68–88. Princeton University Press. https://doi.org/10.1515/9781400860180.68.

Kattge, J., S. Díaz, S. Lavorel, I. C. Prentice, P. Leadley, G. Bönisch, E. Garnier, et al. 2011. "TRY—a Global Database of Plant Traits." *Global Change Biology* 17 (9): 2905–35. https://doi.org/10.1111/j.1365-2486.2011.02451.x.

Kearney, Michael R., Brendan A. Wintle, and Warren P. Porter. 2010. "Correlative and Mechanistic Models of Species Distribution Provide Congruent Forecasts under Climate Change." *Conservation Letters* 3 (3): 203–13. https://doi.org/10.1111/j.1755-263X.2010.00097.x.

Keck, François, Rosetta C. Blackman, Raphael Bossart, Jeanine Brantschen, Marjorie Couton, Samuel Hürlemann, Dominik Kirschner, Nadine Locher, Heng Zhang, and Florian Altermatt. 2022. "Meta-analysis Shows Both Congruence and Complementarity of DNA and eDNA Metabarcoding to Traditional Methods for Biological Community Assessment." *Molecular Ecology* 31 (6): 1820–35. https://doi.org/10.1111/mec.16364.

Keddy, Paul. 1992. "Thoughts on a Review of a Critique for Ecology." *Bulletin of the Ecological Society of America* 73 (4): 234–36.

Keller, Kane R., and Jennifer A. Lau. 2018. "When Mutualisms Matter: Rhizobia Effects on Plant Communities Depend on Host Plant Population and Soil Nitrogen Availability." *Journal of Ecology* 106 (3): 1046–56. https://doi.org/10.1111/1365-2745.12938.

Kendall, Bruce E., Cheryl J. Briggs, William W. Murdoch, Peter Turchin, Stephen P. Ellner, Edward McCauley, Roger M. Nisbet, and Simon N. Wood. 1999. "Why Do Populations Cycle? A Synthesis of Statistical and Mechanistic Modeling Approaches." *Ecology* 80 (6): 1789–1805. https://doi.org/10.2307/176658.

Kendall, Bruce E., John Prendergast, and Ottar N. Bjørnstad. 1998. "The Macroecology of Population Dynamics: Taxonomic and Biogeographic Patterns in Population Cycles." *Ecology Letters* 1 (3): 160–64. https://doi.org/10.1046/j.1461-0248.1998.00037.x.

Kim, Genevieve. 2024. "Faster Than Lightspeed: These Neutrinos Were Faster Than the Speed of Light—Until They Weren't." *Yale Scientific Magazine*, February 11. https://www.yalescientific.org/2024/02/faster-than-lightspeed-these-neutrinos-were-faster-than-the-speed-of-light-until-they-werent/.

Kingsland, Sharon E. 1995. *Modeling Nature*. 2nd ed. Science and Its Conceptual Foundations Series. University of Chicago Press.

———. 2005. *The Evolution of American Ecology, 1890–2000*. JHU Press.

Kingsolver, J. G., H. E. Hoekstra, J. M. Hoekstra, D. Berrigan, S. N. Vignieri, C. E. Hill, A. Hoang, P. Gibert, and P. Beerli. 2001. "The Strength of Phenotypic Selection in Natural Populations." *American Naturalist* 157 (3): 245–61. https://doi.org/10.1086/319193.

Klausmeier, Christopher A. 2010. "Successional State Dynamics: A Novel Approach to Modeling Nonequilibrium Foodweb Dynamics." *Journal of Theoretical Biology* 262 (4): 584–95. https://doi.org/10.1016/j.jtbi.2009.10.018.

Klausmeier, Christopher A., Elena Litchman, Tanguy Daufresne, and Simon A. Levin. 2004. "Optimal Nitrogen-to-Phosphorus Stoichiometry of Phytoplankton." *Nature* 429 (6988): 171–74. https://doi.org/10.1038/nature02454.

Kleiber, M. 1932. "Body Size and Metabolism." *Hilgardia* 6 (11): 315–53.

Knott, Jonathan, Elizabeth LaRue, Samuel Ward, Emily McCallen, Kimberly Ordonez, Franklin Wagner, Insu Jo, Jessica Elliott, and Songlin Fei. 2019. "A Roadmap for Exploring the Thematic Content of Ecology Journals." *Ecosphere* 10 (8): e02801. https://doi.org/10.1002/ecs2.2801.

Knuuttila, Tarja, and Andrea Loettgers. 2011. "Causal Isolation Robustness Analysis: The Combinatorial Strategy of Circadian Clock Research." *Biology and Philosophy* 26 (5): 773–91. https://doi.org/10.1007/s10539-011-9279-x.

Koricheva, Julia. 2003. "Non-significant Results in Ecology: A Burden or a Blessing in Disguise?" *Oikos* 102 (2): 397–401.

Koricheva, Julia, Jessica Gurevitch, and Kerrie Mengersen, eds. 2013. *Handbook of Meta-analysis in Ecology and Evolution*. Princeton University Press.

Krabbenhoft, Corey A., and Donna R. Kashian. 2020. "Citizen Science Data Are a Reliable Complement to Quantitative Ecological Assessments in Urban Rivers." *Ecological Indicators* 116 (September): 106476. https://doi.org/10.1016/j.ecolind.2020.106476.

Kraft, Nathan J. B., Peter B. Adler, Oscar Godoy, Emily C. James, Steve Fuller, and Jonathan M. Levine. 2015. "Community Assembly, Coexistence and the Environmental Filtering Metaphor." *Functional Ecology* 29 (5): 592–99. https://doi.org/10.1111/1365-2435.12345.

Krebs, Charles. 2015a. "A Survey of Strong Inference in Ecology Papers: Platt's Test and Medawar's Fraud Model." *Ecological Rants* (blog), March 10. https://www.zoology.ubc.ca/~krebs/ecological_rants/a-survey-ofstrong-inference-in-ecology-papers-platts-test-and-medawars-fraud-model/.

———. 2015b. "The Volkswagen Syndrome and Ecological Science." *Ecological Rants* (blog), November 7. https://www.zoology.ubc.ca/~krebs/ecological_rants/the-volkswagen-syndrome-and-ecological-science/.

Lai, Jiangshan, Christopher J. Lortie, Robert A. Muenchen, Jian Yang, and Keping Ma. 2019. "Evaluating the Popularity of R in Ecology." *Ecosphere* 10 (1): e02567. https://doi.org/10.1002/ecs2.2567.

Laland, Kevin N., Tobias Uller, Marcus W. Feldman, Kim Sterelny, Gerd B. Müller, Armin Moczek, Eva Jablonka, and John Odling-Smee. 2015. "The Extended Evolutionary Synthesis: Its Structure, Assumptions and Predictions." *Proceedings of the Royal Society B: Biological Sciences* 282 (1813): 20151019. https://doi.org/10.1098/rspb.2015.1019.

Lamy, Thomas, Nathan I. Wisnoski, Riley Andrade, Max C. N. Castorani, Aldo Compagnoni, Nina Lany, Luca Marazzi, et al. 2021. "The Dual Nature of Metacommunity Variability." *Oikos* 130 (12): 2078–92. https://doi.org/10.1111/oik.08517.

Lande, Russell. 1996. "Statistics and Partitioning of Species Diversity, and Similarity among Multiple Communities." *Oikos* 76 (1): 5–13. https://doi.org/10.2307/3545743.

Lawton, John H. 1992. "Predictable Plots." *Nature* 354:444.

———. 1996. "The Ecotron Facility at Silwood Park: The Value of 'Big Bottle' Experiments." *Ecology* 77 (3): 665–69. https://doi.org/10.2307/2265488.

———. 1999. "Are There General Laws in Ecology?" *Oikos* 84 (2): 177–92. https://doi.org/10.2307/3546712.

Leibold, M. A., M. Holyoak, N. Mouquet, P. Amarasekare, J. M. Chase, M. F. Hoopes, R. D. Holt, et al. 2004. "The Metacommunity Concept: A

Framework for Multi-scale Community Ecology." *Ecology Letters* 7 (7): 601–13. https://doi.org/10.1111/j.1461-0248.2004.00608.x.

Lemoine, Nathan P., Ava Hoffman, Andrew J. Felton, Lauren Baur, Francis Chaves, Jesse Gray, Qiang Yu, and Melinda D. Smith. 2016. "Underappreciated Problems of Low Replication in Ecological Field Studies." *Ecology* 97 (10): 2554–61. https://doi.org/10.1002/ecy.1506.

Lemoine, Rhys Taylor, Robert Buitenwerf, and Jens-Christian Svenning. 2023. "Megafauna Extinctions in the Late-Quaternary Are Linked to Human Range Expansion, Not Climate Change." *Anthropocene* 44 (December): 100403. https://doi.org/10.1016/j.ancene.2023.100403.

Leslie, Mitchell. 2000. "Divided They Stand." *Stanford Magazine*, January/February. https://stanfordmag.org/contents/divided-they-stand.

Levins, Richard. 1966. "The Strategy of Model Building in Population Biology." *American Scientist* 54 (4): 421–31.

Lewis, Abigail S. L., Whitney M. Woelmer, Heather L. Wander, Dexter W. Howard, John W. Smith, Ryan P. McClure, Mary E. Lofton, et al. 2022. "Increased Adoption of Best Practices in Ecological Forecasting Enables Comparisons of Forecastability." *Ecological Applications* 32 (2): e2500. https://doi.org/10.1002/eap.2500.

Liebhold, Andrew, Walter D. Koenig, and Ottar N. Bjørnstad. 2004. "Spatial Synchrony in Population Dynamics." *Annual Review of Ecology, Evolution, and Systematics* 35 (1): 467–90. https://doi.org/10.1146/annurev.ecolsys.34.011802.132516.

Lindenmayer, David B., and Joern Fischer. 2007. "Tackling the Habitat Fragmentation Panchreston." *Trends in Ecology & Evolution* 22 (3): 127–32. https://doi.org/10.1016/j.tree.2006.11.006.

Lindenmayer, David B., and Gene E. Likens. 2011. "Losing the Culture of Ecology." *Bulletin of the Ecological Society of America* 92 (3): 245–46. https://doi.org/10.1890/0012-9623-92.3.245.

———. 2013. "Benchmarking Open Access Science Against Good Science." *Bulletin of the Ecological Society of America* 94 (4): 338–40. https://doi.org/10.1890/0012-9623-94.4.338.

Linquist, Stefan, T. Ryan Gregory, Tyler A. Elliott, Brent Saylor, Stefan C. Kremer, and Karl Cottenie. 2016. "Yes! There Are Resilient Generalizations (or 'Laws') in Ecology." *Quarterly Review of Biology* 91 (2): 119–31. https://doi.org/10.1086/686809.

Loehle, Craig. 1987. "Hypothesis Testing in Ecology: Psychological Aspects and the Importance of Theory Maturation." *Quarterly Review of Biology* 62 (4): 397–409. https://doi.org/10.1086/415619.

———. 2011. "Complexity and the Problem of Ill-Posed Questions in Ecology." *Ecological Complexity* 8 (1): 60–67. https://doi.org/10.1016/j.ecocom.2010.11.004.

Logue, Jürg B., Nicolas Mouquet, Hannes Peter, and Helmut Hillebrand. 2011. "Empirical Approaches to Metacommunities: A Review and Comparison with Theory." *Trends in Ecology & Evolution* 26 (9): 482–91. https://doi.org/10.1016/j.tree.2011.04.009.

Loreau, Michel, and Claire de Mazancourt. 2008. "Species Synchrony and Its Drivers: Neutral and Nonneutral Community Dynamics in Fluctuating Environments." *American Naturalist* 172 (2): E48–E66. https://doi.org/10.1086/589746.

Loreau, Michel, and Andy Hector. 2001. "Partitioning Selection and Complementarity in Biodiversity Experiments." *Nature* 412 (6842): 72–76. https://doi.org/10.1038/35083573.

Losos, Jonathan B. 2009. *Lizards in an Evolutionary Tree: Ecology and Adaptive Radiation of Anoles*. University of California Press.

Low-Décarie, Etienne, Corey Chivers, and Monica Granados. 2014. "Rising Complexity and Falling Explanatory Power in Ecology." *Frontiers in Ecology and the Environment* 12 (7): 412–18. https://doi.org/10.1890/130230.

Luckinbill, Leo S. 1973. "Coexistence in Laboratory Populations of *Paramecium aurelia* and Its Predator *Didinium nasutum*." *Ecology* 54 (6): 1320–27. https://doi.org/10.2307/1934194.

MacArthur, Robert H. 1957. "On the Relative Abundance of Bird Species." *Proceedings of the National Academy of Sciences* 43 (3): 293–95. https://doi.org/10.1073/pnas.43.3.293.

———. 1972. *Geographical Ecology: Patterns in the Distribution of Species*. Princeton University Press.

MacArthur, Robert H., and Richard Levins. 1967. "The Limiting Similarity, Convergence, and Divergence of Coexisting Species." *American Naturalist* 101 (921): 377–85.

MacArthur, Robert H., and Edward O. Wilson. 1963. "An Equilibrium Theory of Insular Zoogeography." *Evolution* 17 (4): 373–87. https://doi.org/10.1111/j.1558-5646.1963.tb03295.x.

———. 1967. *The Theory of Island Biogeography*. Rev. ed. Princeton University Press.

Maier, Donald S. 2012. *What's So Good about Biodiversity? A Call for Better Reasoning about Nature's Value*. Springer.

Maiorana, Virginia C. 1978. "An Explanation of Ecological and Developmental Constants." *Nature* 273 (5661): 375–77. https://doi.org/10.1038/273375a0.

Margulis, Lynn. 1991. *Symbiosis as a Source of Evolutionary Innovation: Speciation and Morphogenesis*. Edited by Rene Fester. MIT Press.

Mark, Robert. 1996. "Architecture and Evolution." *American Scientist* 84 (4): 383–89.

Marquet, Pablo A., Andrew P. Allen, James H. Brown, Jennifer A. Dunne, Brian J. Enquist, James F. Gillooly, Patricia A. Gowaty, et al. 2014. "On Theory in Ecology." *BioScience* 64 (8): 701–10. https://doi.org/10.1093/biosci/biu098.

Martin, Laura J, Bernd Blossey, and Erle Ellis. 2012. "Mapping Where Ecologists Work: Biases in the Global Distribution of Terrestrial Ecological Observations." *Frontiers in Ecology and the Environment* 10 (4): 195–201. https://doi.org/10.1890/110154.

Maurer, Brian A. 1999. *Untangling Ecological Complexity: The Macroscopic Perspective*. 1st ed. University of Chicago Press.

May, Robert M. 1972. "Will a Large Complex System Be Stable?" *Nature* 238 (5364): 413–14. https://doi.org/10.1038/238413a0.

Mayfield, Margaret M., and Jonathan M. Levine. 2010. "Opposing Effects of Competitive Exclusion on the Phylogenetic Structure of Communities." *Ecology Letters* 13 (9): 1085–93. https://doi.org/10.1111/j.1461-0248.2010.01509.x.

McCallen, Emily, Jonathan Knott, Gabriela Nunez-Mir, Benjamin Taylor, Insu Jo, and Songlin Fei. 2019. "Trends in Ecology: Shifts in Ecological Research Themes over the Past Four Decades." *Frontiers in Ecology and the Environment* 17 (2): 109–16. https://doi.org/10.1002/fee.1993.

McCann, Kevin S. 2000. "The Diversity-Stability Debate." *Nature* 405 (6783): 228–33. https://doi.org/10.1038/35012234.

McCann, Kevin S., Alan Hastings, and Gary R. Huxel. 1998. "Weak Trophic Interactions and the Balance of Nature." *Nature* 395 (6704): 794–98. https://doi.org/10.1038/27427.

McCauley, Edward, Roger M. Nisbet, William W. Murdoch, Andre M. de Roos, and William S. C. Gurney. 1999. "Large-Amplitude Cycles of *Daphnia* and Its Algal Prey in Enriched Environments." *Nature* 402 (6762): 653–56. https://doi.org/10.1038/45223.

McGill, Brian J. 2003. "A Test of the Unified Neutral Theory of Biodiversity." *Nature* 422 (6934): 881–85. https://doi.org/10.1038/nature01583.

———. 2014. "Scientists Have to Present a United Front, Right?" *Dynamic Ecology* (blog), April 21. https://dynamicecology.wordpress.com/2014/04/21/scientists-have-to-present-a-united-front-right/.

———. 2015. "Why AIC Appeals to Ecologist's Lowest Instincts." *Dynamic Ecology* (blog), May 21. https://dynamicecology.wordpress.com/2015/05/21/why-aic-appeals-to-ecologists-lowest-instincts/.

———. 2024. "The State of Academic Publishing in 3 Graphs, 6 Trends, and 4 Thoughts." *Dynamic Ecology* (blog), April 29. https://dynamicecology.wordpress.com/2024/04/29/the-state-of-academic-publishing-in-3-graphs-5-trends-and-4-thoughts/.

McGill, Brian J., Rampal S. Etienne, John S. Gray, David Alonso, Marti J. Anderson, Habtamu Kassa Benecha, Maria Dornelas, et al. 2007. "Species Abundance Distributions: Moving beyond Single Prediction Theories to Integration within an Ecological Framework." *Ecology Letters* 10 (10): 995–1015. https://doi.org/10.1111/j.1461-0248.2007.01094.x.

Menzies, Allyson. 2020. "About Me." http://allysonmenzies.weebly.com/about-me.html. Accessed July 2022.

Mertz, David B., and David E. McCauley. 1980. "The Domain of Laboratory Ecology." *Synthese* 43 (1): 95–110. https://doi.org/10.1007/BF00413858.

Meyer, Carsten, Patrick Weigelt, and Holger Kreft. 2016. "Multidimensional Biases, Gaps and Uncertainties in Global Plant Occurrence Information." *Ecology Letters* 19 (8): 992–1006. https://doi.org/10.1111/ele.12624.

Michener, William K. 2015. "Ecological Data Sharing." *Ecological Informatics* 29 (September): 33–44. https://doi.org/10.1016/j.ecoinf.2015.06.010.

Miller, Richard S., and John F. Reed. 1965. "Summary Report of the Ecology Study Committee with Recommendations for the Future of Ecology and the Ecological Society of America." *Bulletin of the Ecological Society of America* 46 (2): 61–82.

Milo, Ron, Shalev Itzkovitz, Nadav Kashtan, Reuven Levitt, Shai Shen-Orr, Inbal Ayzenshtat, Michal Sheffer, and Uri Alon. 2004. "Superfamilies of Evolved and Designed Networks." *Science* 303 (5663): 1538–42. https://doi.org/10.1126/science.1089167.

Milo, Ron, S. Shen-Orr, S. Itzkovitz, N. Kashtan, D. Chklovskii, and U. Alon. 2002. "Network Motifs: Simple Building Blocks of Complex Networks." *Science* 298 (5594): 824–27. https://doi.org/10.1126/science.298.5594.824.

Moles, Angela T., Habacuc Flores-Moreno, Stephen P. Bonser, David I. Warton, Aveliina Helm, Laura Warman, David J. Eldridge, et al. 2012. "Invasions: The Trail Behind, the Path Ahead, and a Test of a Disturbing Idea." *Journal of Ecology* 100 (1): 116–27. https://doi.org/10.1111/j.1365-2745.2011.01915.x.

Moran, Nicholas P., Alfredo Sánchez-Tójar, Holger Schielzeth, and Klaus Reinhold. 2021. "Poor Nutritional Condition Promotes High-Risk Behaviours: A Systematic Review and Meta-analysis." *Biological Reviews* 96 (1): 269–88. https://doi.org/10.1111/brv.12655.

Morin, Xavier, and Wilfried Thuiller. 2009. "Comparing Niche- and Process-Based Models to Reduce Prediction Uncertainty in Species Range Shifts under Climate Change." *Ecology* 90 (5): 1301–13. https://doi.org/10.1890/08-0134.1.

Mott, Rowan, and Rohan H. Clarke. 2018. "Systematic Review of Geographic Biases in the Collection of At-Sea Distribution Data for Seabirds." *Emu—Austral Ornithology* 118 (3): 235–46. https://doi.org/10.1080/01584197.2017.1416957.

Mountz, Alison, and Kira Williams. 2023. "Let Geography Die: The Rise, Fall, and 'Unfinished Business' of Geography at Harvard." *Annals of the American Association of Geographers* 113 (8): 1977–2002. https://doi.org/10.1080/24694452.2023.2208645.

Munday, Philip L., Danielle L. Dixson, Jennifer M. Donelson, Geoffrey P. Jones, Morgan S. Pratchett, Galina V. Devitsina, and Kjell B. Døving. 2009. "Ocean Acidification Impairs Olfactory Discrimination and Homing Ability of a Marine Fish." *Proceedings of the National Academy of Sciences* 106 (6): 1848–52. https://doi.org/10.1073/pnas.0809996106.

Munday, Philip L., Danielle L. Dixson, Mark I. McCormick, Mark Meekan, Maud C. O. Ferrari, and Douglas P. Chivers. 2010. "Replenishment of Fish Populations Is Threatened by Ocean Acidification." *Proceedings of the National Academy of Sciences* 107 (29): 12930–34. https://doi.org/10.1073/pnas.1004519107.

Murdoch, W. W., B. E. Kendall, R. M. Nisbet, C. J. Briggs, E. McCauley, and R. Bolser. 2002. "Single-Species Models for Many-Species Food Webs." *Nature* 417 (6888): 541–43. https://doi.org/10.1038/417541a.

Murray, Bertram G. 2000. "Universal Laws and Predictive Theory in Ecology and Evolution." *Oikos* 89 (2): 403–8. https://doi.org/10.1034/j.1600-0706.2000.890223.x.

Murray, James D. 2001. *An Introduction to Mathematical Biology*. 3rd ed. Springer.

Nelson, Gil, and Shari Ellis. 2018. "The History and Impact of Digitization and Digital Data Mobilization on Biodiversity Research." *Philosophical Transactions of the Royal Society B: Biological Sciences* 374 (1763): 20170391. https://doi.org/10.1098/rstb.2017.0391.

Nilsen, Erlend B., Diana E. Bowler, and John D. C. Linnell. 2020. "Exploratory and Confirmatory Research in the Open Science Era." *Journal of Applied Ecology* 57 (4): 842–47. https://doi.org/10.1111/1365-2664.13571.

NOAA Fisheries. 2023. "90-Day Finding on a Petition to List the Tope Shark under the Endangered Species Act." July 25. https://www.fisheries.noaa.gov/action/90-day-finding-petition-list-tope-shark-under-endangered-species-act.

Noble, Andrew E., Todd S. Rosenstock, Patrick H. Brown, Jonathan Machta, and Alan Hastings. 2018. "Spatial Patterns of Tree Yield Explained by Endogenous Forces through a Correspondence between the Ising Model and Ecology." *Proceedings of the National Academy of Sciences* 115 (8): 1825–30. https://doi.org/10.1073/pnas.1618887115.

Noble, Denis. 2013. "Physiology Is Rocking the Foundations of Evolutionary Biology." *Experimental Physiology* 98 (8): 1235–43. https://doi.org/10.1113/expphysiol.2012.071134.

NSF National Science Foundation. n.d.-a. "History—About NSF." Accessed October 24, 2024. https://new.nsf.gov/about/history.

———. n.d.-b. "Publications Output: U.S. Trends and International Comparisons." Accessed October 24, 2024. https://ncses.nsf.gov/pubs/nsb202333/publication-output-by-region-country-or-economy-and-by-scientific-field.

Nuñez, Martin A., Jos Barlow, Marc Cadotte, Kirsty Lucas, Erika Newton, Nathalie Pettorelli, and Philip A. Stephens. 2019. "Assessing the Uneven Global Distribution of Readership, Submissions and Publications in Applied Ecology: Obvious Problems without Obvious Solutions." *Journal of Applied Ecology* 56 (1): 4–9. https://doi.org/10.1111/1365-2664.13319.

O'Brien, Allyson, Kallie Townsend, Robin Hale, David Sharley, and Vincent Pettigrove. 2016. "How Is Ecosystem Health Defined and Measured? A Critical Review of Freshwater and Estuarine Studies." *Ecological Indicators* 69 (October): 722–29. https://doi.org/10.1016/j.ecolind.2016.05.004.

O'Connor, Mary I., Johnna M. Holding, Carrie V. Kappel, Carlos M. Duarte, Keith Brander, Christopher J. Brown, John F. Bruno, et al. 2015. "Strengthening Confidence in Climate Change Impact Science." *Global Ecology and Biogeography* 24 (1): 64–76. https://doi.org/10.1111/geb.12218.

Odenbaugh, Jay. 2005. "Idealized, Inaccurate but Successful: A Pragmatic Approach to Evaluating Models in Theoretical Ecology." *Biology and Philosophy* 20 (2–3): 231–55. https://doi.org/10.1007/s10539-004-0478-6.

———. 2011. "True Lies: Realism, Robustness, and Models." *Philosophy of Science* 78 (5): 1177–88. https://doi.org/10.1086/662281.

Odenbaugh, Jay, and Anna Alexandrova. 2011. "Buyer Beware: Robustness Analyses in Economics and Biology." *Biology and Philosophy* 26 (5): 757–71. https://doi.org/10.1007/s10539-011-9278-y.

Okasha, Samir. 2004. "Multilevel Selection and the Partitioning of Covariance: A Comparison of Three Approaches." *Evolution* 58 (3): 486–94. https://doi.org/10.1111/j.0014-3820.2004.tb01672.x.

Oksanen, Lauri, Stephen D. Fretwell, Joseph Arruda, and Pekka Niemela. 1981. "Exploitation Ecosystems in Gradients of Primary Productivity." *American Naturalist* 118 (2): 240–61.

Ollerton, Jeff. 2015. "Evolving a Naturalist—Happy Birthday to Me!" *Prof. Jeff Ollerton—Ecological Scientist and Author* (blog), February 11. https://jeffollerton.co.uk/2015/02/11/evolving-a-naturalist/.

O'Malley, Maureen A. 2015. "Endosymbiosis and Its Implications for Evolutionary Theory." *Proceedings of the National Academy of Sciences* 112 (33): 10270–77. https://doi.org/10.1073/pnas.1421389112.

Onorato, Dave P., Mark W. Cunningham, Mark Lotz, Marc Criffield, David Shindle, Annette Johnson, Bambi C. F. Clemons, et al. 2024. "Multi-generational Benefits of Genetic Rescue." *Scientific Reports* 14 (1): 17519. https://doi.org/10.1038/s41598-024-67033-6.

Oransky, Ivan. 2011. "A Flying What? Symbiosis Retracts Paper Claiming New Species Arise from Accidental Mating." *Retraction Watch* (blog), November 2. https://retractionwatch.com/2011/11/02/a-flying-what-symbiosis-retracts-paper-claiming-new-species-arise-from-accidental-mating/.

Oreskes, Naomi. 1999. *The Rejection of Continental Drift: Theory and Method in American Earth Science*. Illus. ed. Oxford University Press.

O'Rourke, Kevin. 2006. "A Historical Perspective on Meta-analysis: Dealing Quantitatively with Varying Study Results." James Lind Library. Accessed in 2023. https://www.jameslindlibrary.org/articles/a-historical-perspective-on-meta-analysis-dealing-quantitatively-with-varying-study-results/.

Orzack, Steven Hecht, and Elliott Sober. 1993. "A Critical Assessment of Levins's 'The Strategy of Model Building in Population Biology' (1966)." *Quarterly Review of Biology* 68 (4): 533–46.

Otto, Sarah P., and Alirio Rosales. 2020. "Theory in Service of Narratives in Evolution and Ecology." *American Naturalist* 195 (2): 290–99. https://doi.org/10.1086/705991.

Ou, William J.-A., Gil J. B. Henriques, Athmanathan Senthilnathan, Po-Ju Ke, Tess N. Grainger, and Rachel M. Germain. 2022. "Writing Accessible Theory in Ecology and Evolution: Insights from Cognitive Load Theory." *BioScience* 72 (3): 300–313. https://doi.org/10.1093/biosci/biab133.

Paine, C. E. Timothy, and Charles W. Fox. 2018. "The Effectiveness of Journals as Arbiters of Scientific Impact." *Ecology and Evolution* 8 (19): 9566–85.

Paine, Robert T. 2010. "Macroecology: Does It Ignore or Can It Encourage Further Ecological Syntheses Based on Spatially Local Experimental Manipulations?" *American Naturalist* 176 (4): 385–93. https://doi.org/10.1086/656273.

Palmer, Michael W. 2007. "Species–Area Curves and the Geometry of Nature." In *Scaling Biodiversity*, edited by David Storch, James Brown, and Pablo Marquet, 15–31. Ecological Reviews. Cambridge University Press. https://doi.org/10.1017/CBO9780511814938.004.

Parker, Wendy S. 2011. "When Climate Models Agree: The Significance of Robust Model Predictions*." *Philosophy of Science* 78 (4): 579–600. https://doi.org/10.1086/661566.

Parris, Kirsten, and Michael McCarthy. 2001. "Identifying Effects of Toe Clipping on Anuran Return Rates: The Importance of Statistical Power." *Amphibia-Reptilia* 22 (3): 275–89.

Pauly, D., and R. S. V. Pullin. 1998. "Hatching Time in Spherical, Pelagic, Marine Fish Eggs in Response to Temperature and Egg Size." *Environmental Biology of Fishes* 5:261–71. https://doi.org/10.1007/BF00004892.

Payne-Gaposchkin, Cecilia. 1925. "Stellar Atmospheres; a Contribution to the Observational Study of High Temperature in the Reversing Layers of Stars." PhD diss., Harvard College.

Peters, Douglas P., and Stephen J. Ceci. 1982. "Peer-Review Practices of Psychological Journals: The Fate of Published Articles, Submitted Again." *Behavioral and Brain Sciences* 5 (2): 187–95. https://doi.org/10.1017/S0140525X00011183.

Peters, Robert Henry. 1983. *The Ecological Implications of Body Size*. Cambridge Studies in Ecology. Cambridge University Press.

———. 1991. *A Critique for Ecology*. Illus. ed. Cambridge University Press.

Pettorelli, Nathalie, Jos Barlow, Martin A. Nuñez, Romina Rader, Philip A. Stephens, Thomas Pinfield, and Erika Newton. 2021. "How International Journals Can Support Ecology from the Global South." *Journal of Applied Ecology* 58 (1): 4–8. https://doi.org/10.1111/1365-2664.13815.

Pfab, Thiemo, Torsten Slowinski, Michael Godes, Horst Halle, Friedrich Priem, and Berthold Hocher. 2006. "Low Birth Weight, a Risk Factor for Cardiovascular Diseases in Later Life, Is Already Associated with Elevated Fetal Glycosylated Hemoglobin at Birth." *Circulation* 114:1687–92. https://doi.org/10.1161/CIRCULATIONAHA.106.625848.

Pianka, Eric R., Laurie J. Vitt, Nicolás Pelegrin, Daniel B. Fitzgerald, and Kirk O. Winemiller. 2017. "Toward a Periodic Table of Niches, or Exploring the Lizard Niche Hypervolume." *American Naturalist* 190 (5): 601–16. https://doi.org/10.1086/693781.

Pironon, S., I. Ondo, M. Diazgranados, R. Allkin, A. C. Baquero, R. Cámara-Leret, C. Canteiro, et al. 2024. "The Global Distribution of Plants Used by Humans." *Science* 383 (6680): 293–97. https://doi.org/10.1126/science.adg8028.

Pitman, Nigel C. A., Jocelyn Widmer, Clinton N. Jenkins, Gabriela Stocks, Lisa Seales, Franklin Paniagua, and Emilio M. Bruna. 2011. "Volume and Geographical Distribution of Ecological Research in the Andes and the Amazon, 1995–2008." *Tropical Conservation Science* 4 (1): 64–81. https://doi.org/10.1177/194008291100400107.

Platt, John R. 1964. "Strong Inference." *Science* 146 (3642): 347–53. https://doi.org/10.1126/science.146.3642.347.

Pocock, Michael J. O., John C. Tweddle, Joanna Savage, Lucy D. Robinson, and Helen E. Roy. 2017. "The Diversity and Evolution of Ecological and Environmental Citizen Science." *PLOS ONE* 12 (4): e0172579. https://doi.org/10.1371/journal.pone.0172579.

Pokallus, J. W., G. M. Campbell, B. J. Koch, and J. N. Pauli. 2011. "The Landscape of Ecology." *Ecosphere* 2 (2): art22. https://doi.org/10.1890/ES10-00173.1.

Popovic, Gordana, Tanya Jane Mason, Szymon Marian Drobniak, Tiago André Marques, Joanne Potts, Rocío Joo, Res Altwegg, et al. 2024. "Four Principles for Improved Statistical Ecology." *Methods in Ecology and Evolution* 15 (2): 266–81. https://doi.org/10.1111/2041-210X.14270.

Powers, Jennifer S., Marife D. Corre, Tracy E. Twine, Edzo Veldkamp, and Susan E. Trumbore. 2011. "Geographic Bias of Field Observations of Soil Carbon Stocks with Tropical Land-Use Changes Precludes Spatial Extrapolation." *Proceedings of the National Academy of Sciences* 108 (15): 6318–22. https://doi.org/10.1073/pnas.1016774108.

Price, George R. 1970. "Selection and Covariance." *Nature* 227 (5257): 520–21. https://doi.org/10.1038/227520a0.

———. 1972. "Extension of Covariance Selection Mathematics." *Annals of Human Genetics* 35 (4): 485–90. https://doi.org/10.1111/j.1469-1809.1957.tb01874.x.

Primack, Richard B., Tara K. Miller, Carina Terry, Erika Marín-Spiotta, Pamela H. Templer, Asmeret Asefaw Berhe, Emily J. Diaz Vallejo, et al. 2023. "Historically Excluded Groups in Ecology Are Undervalued and Poorly Treated." *Frontiers in Ecology and the Environment* 21 (8): 363–69. https://doi.org/10.1002/fee.2613.

Pyke, Graham H. 2015. "Understanding Movements of Organisms: It's Time to Abandon the Lévy Foraging Hypothesis." *Methods in Ecology and Evolution* 6 (1): 1–16. https://doi.org/10.1111/2041-210X.12298.

Pyšek, Petr, David M. Richardson, Jan Pergl, Vojtech Jarosík, Zuzana Sixtová, and Ewald Weber. 2008. "Geographical and Taxonomic Biases in Invasion Ecology." *Trends in Ecology & Evolution* 23 (5): 237–44. https://doi.org/10.1016/j.tree.2008.02.002.

Queller, David C. 1995. "The Spaniels of St. Marx and the Panglossian Paradox: A Critique of a Rhetorical Programme." *Quarterly Review of Biology* 70 (4): 485–89. https://doi.org/10.1086/419174.

Quinn, James F., and Arthur E. Dunham. 1983. "On Hypothesis Testing in Ecology and Evolution." *American Naturalist* 122 (5): 602–17.

Rangel, Thiago F. L. V. B., and José A. F. Diniz-Filho. 2005. "Neutral Community Dynamics, the Mid-domain Effect and Spatial Patterns in Species Richness." *Ecology Letters* 8 (8): 783–90. https://doi.org/10.1111/j.1461-0248.2005.00786.x.

Rankin, Brian D., Jeremy W. Fox, Christian R. Barrón-Ortiz, Amy E. Chew, Patricia A. Holroyd, Joshua A. Ludtke, Xingkai Yang, and Jessica M. Theodor. 2015. "The Extended Price Equation Quantifies Species Selection on Mammalian Body Size across the Palaeocene/Eocene Thermal Maximum." *Proceedings of the Royal Society B: Biological Sciences* 282 (1812): 20151097. https://doi.org/10.1098/rspb.2015.1097.

Rastogi, Charvi, Ivan Stelmakh, Alina Beygelzimer, Yann N. Dauphin, Percy Liang, Jennifer Wortman Vaughan, Zhenyu Xue, Hal Daumé III, Emma Pierson, and Nihar B. Shah. 2024. "How Do Authors' Perceptions of Their Papers Compare with Co-authors' Perceptions and Peer-Review Decisions?" *PLOS ONE* 19 (4): e0300710. https://doi.org/10.1371/journal.pone.0300710.

Reddy, Sushma, and Liliana M. Dávalos. 2003. "Geographical Sampling Bias and Its Implications for Conservation Priorities in Africa." *Journal of Biogeography* 30 (11): 1719–27. https://doi.org/10.1046/j.1365-2699.2003.00946.x.

Ricklefs, Robert E. 1987. "Community Diversity: Relative Roles of Local and Regional Processes." *Science* 235 (4785): 167–71. https://doi.org/10.1126/science.235.4785.167.

Ricotta, Carlo. 2017. "Of Beta Diversity, Variance, Evenness, and Dissimilarity." *Ecology and Evolution* 7 (13): 4835–43. https://doi.org/10.1002/ece3.2980.

Ricotta, Carlo, and Enrico Feoli. 2024. "Hill Numbers Everywhere. Does It Make Ecological Sense?" *Ecological Indicators* 161 (April): 111971. https://doi.org/10.1016/j.ecolind.2024.111971.

Ritchie, Hannah, Lucas Rodés-Guirao, Edouard Mathieu, Marcel Gerber, Esteban Ortiz-Ospina, Joe Hasell, and Max Roser. 2023. "Population Growth." Our World in Data. Accessed June 2024. https://ourworldindata.org/population-growth.

Riva, Federico, Nicola Koper, and Lenore Fahrig. 2024. "Overcoming Confusion and Stigma in Habitat Fragmentation Research." *Biological Reviews* 99 (4): 1411–24. https://doi.org/10.1111/brv.13073.

Roche, Dominique G., Ilias Berberi, Fares Dhane, Félix Lauzon, Sandrine Soeharjono, Roslyn Dakin, and Sandra A. Binning. 2022. "Slow Improvement to the Archiving Quality of Open Datasets Shared by Researchers in Ecology and Evolution." *Proceedings of the Royal Society B: Biological Sciences* 289 (1975): 20212780. https://doi.org/10.1098/rspb.2021.2780.

Rosenberg, M. S., D. C. Adams, and Jessica Gurevitch. 1997. MetaWin (software package). Sinauer Associates, Sunderland, Massachusetts.

Rosenzweig, Michael L. 1995. *Species Diversity in Space and Time.* Cambridge University Press.

Rosenzweig, M. L., and R. H. MacArthur. 1963. "Graphical Representation and Stability Conditions of Predator-Prey Interactions." *American Naturalist* 97 (895): 209–23.

Rossberg, Axel G., György Barabás, Hugh P. Possingham, Mercedes Pascual, Pablo A. Marquet, Cang Hui, Matthew R. Evans, and Géza Meszéna. 2019. "Let's Train More Theoretical Ecologists—Here Is Why." *Trends in Ecology & Evolution* 34 (9): 759–62. https://doi.org/10.1016/j.tree.2019.06.004.

Roth, V. Louise. 1981. "Constancy in the Size Ratios of Sympatric Species." *American Naturalist* 118 (3): 394–404. https://doi.org/10.1086/283831.

Rothwell, Peter M., and Christopher N. Martyn. 2000. "Reproducibility of Peer Review in Clinical Neuroscience: Is Agreement between Reviewers Any Greater Than Would Be Expected by Chance Alone?" *Brain* 123 (9): 1964–69. https://doi.org/10.1093/brain/123.9.1964.

Roughgarden, Jonathan. 1995. *Anolis Lizards of the Caribbean: Ecology, Evolution, and Plate Tectonics*. Oxford Series in Ecology and Evolution. Oxford University Press.

Rovere, Jacob, and Jeremy W. Fox. 2019. "Persistently Rare Species Experience Stronger Negative Frequency Dependence Than Common Species: A Statistical Attractor That Is Hard to Avoid." *Global Ecology and Biogeography* 28 (4): 508–20. https://doi.org/10.1111/geb.12871.

Rubenstein, Madeleine A., Sarah R. Weiskopf, Romain Bertrand, Shawn L. Carter, Lise Comte, Mitchell J. Eaton, Ciara G. Johnson, et al. 2023. "Climate Change and the Global Redistribution of Biodiversity: Substantial Variation in Empirical Support for Expected Range Shifts." *Environmental Evidence* 12 (1): 7. https://doi.org/10.1186/s13750-023-00296-0.

Runge, Claire A., Tara G. Martin, Hugh P. Possingham, Stephen G. Willis, and Richard A. Fuller. 2014. "Conserving Mobile Species." *Frontiers in Ecology and the Environment* 12 (7): 395–402. https://doi.org/10.1890/130237.

Sacks, Benjamin. 2015. "What Happened to the American Geography Department?" *Geography Directions* (blog), April 8. https://blog.geographydirections.com/2015/04/08/what-happened-to-the-american-geography-department/.

Sales, Lilian, Matt Hayward, and Rafael Loyola. 2021. "What Do You Mean by 'Niche'? Modern Ecological Theories Are Not Coherent on Rhetoric about the Niche Concept." *Acta Oecologica* 110 (May): 103701. https://doi.org/10.1016/j.actao.2020.103701.

Salomon, Anne K., Nick M. Tanape Sr., and Henry P. Huntington. 2007. "Serial Depletion of Marine Invertebrates Leads to the Decline of a Strongly Interacting Grazer." *Ecological Applications* 17 (6): 1752–70. https://doi.org/10.1890/06-1369.1.

Sánchez-Ochoa, Daniel, Edgar J. González, Maria Del Coro Arizmendi, Patricia Koleff, Raúl Martell-Dubois, Jorge A. Meave, and Hibraim Adán Pérez-Mendoza. 2022. "Quantifying Phenological Diversity: A Framework Based on Hill Numbers Theory." *PeerJ* 10: e13412. https://doi.org/10.7717/peerj.13412.

Santayana, G. 1905. *The Life of Reason*. Charles Scribner's Sons.

Saunders, Manu E., Meghan A. Duffy, Stephen B. Heard, Margaret Kosmala, Simon R. Leather, Terrence P. McGlynn, Jeff Ollerton, and Amy L. Parachnowitsch. 2017. "Bringing Ecology Blogging into the Scientific Fold: Measuring Reach and Impact of Science Community Blogs." *Royal Society Open Science* 4 (10): 170957. https://doi.org/10.1098/rsos.170957.

Scheffer, Marten. 2009. *Critical Transitions in Nature and Society*. Illus. ed. Princeton University Press.

Scheiner, Samuel M. 2013. "The Ecological Literature, an Idea-Free Distribution." *Ecology Letters* 16 (12): 1421–23. https://doi.org/10.1111/ele.12196.

Scheiner, Samuel M., and Michael R. Willig. 2008. "A General Theory of Ecology." *Theoretical Ecology* 1 (1): 21–28. https://doi.org/10.1007/s12080-007-0002-0.

———, eds. 2011. *The Theory of Ecology*. University of Chicago Press. https://press.uchicago.edu/ucp/books/book/chicago/T/bo11161054.html.

Schluter, Dolph, and John Donald McPhail. 1992. "Ecological Character Displacement and Speciation in Sticklebacks." *American Naturalist* 140 (1): 85–108.

Schoener, Thomas W. 1983. "Field Experiments on Interspecific Competition." *American Naturalist* 122 (2): 240–85.

Secord, Ross, Jonathan I. Bloch, Stephen G. B. Chester, Doug M. Boyer, Aaron R. Wood, Scott L. Wing, Mary J. Kraus, Francesca A. McInerney, and John Krigbaum. 2012. "Evolution of the Earliest Horses Driven by Climate Change in the Paleocene-Eocene Thermal Maximum." *Science* 335 (6071): 959–62. https://doi.org/10.1126/science.1213859.

Seebens, Hanno, Nicole Schwartz, Peter J. Schupp, and Bernd Blasius. 2016. "Predicting the Spread of Marine Species Introduced by Global Shipping." *Proceedings of the National Academy of Sciences* 113 (20): 5646–51. https://doi.org/10.1073/pnas.1524427113.

Seilacher, Adolf. 1989. "Vendozoa: Organismic Construction in the Proterozoic Biosphere." *Lethaia* 22 (3): 229–39. https://doi.org/10.1111/j.1502-3931.1989.tb01332.x.

———. 1992. "Vendobionta and Psammocorallia: Lost Constructions of Precambrian Evolution." *Journal of the Geological Society* 149 (4): 607–13.

Sells, Sarah N., Sarah B. Bassing, Kristin J. Barker, Shannon C. Forshee, Allison Keever, James W. Goerz, and Michael S. Mitchell. 2018. "Increased Scientific Rigor Will Improve Reliability of Research and Effectiveness of Management." *Journal of Wildlife Management* 82 (3): 485–94. https://doi.org/10.1002/jwmg.21413.

Sender, Ron, Shai Fuchs, and Ron Milo. 2016. "Revised Estimates for the Number of Human and Bacteria Cells in the Body." *PLOS Biology* 14 (8): e1002533. https://doi.org/10.1371/journal.pbio.1002533.

Senior, Alistair M., Catherine E. Grueber, Tsukushi Kamiya, Malgorzata Lagisz, Katie O'Dwyer, Eduardo S. A. Santos, and Shinichi Nakagawa. 2016. "Heterogeneity in Ecological and Evolutionary Meta-analyses: Its Magnitude and Implications." *Ecology* 97 (12): 3293–99. https://doi.org/10.1002/ecy.1591.

Servedio, Maria R. 2020. "An Effective Mutualism? The Role of Theoretical Studies in Ecology and Evolution." *American Naturalist* 195 (2): 284–89. https://doi.org/10.1086/706814.

Servedio, Maria R., Yaniv Brandvain, Sumit Dhole, Courtney L. Fitzpatrick, Emma E. Goldberg, Caitlin A. Stern, Jeremy Van Cleve, and D. Justin Yeh. 2014. "Not Just a Theory—the Utility of Mathematical Models in Evolutionary Biology." *PLOS Biology* 12 (12): e1002017. https://doi.org/10.1371/journal.pbio.1002017.

Shafer, Craig L. 1995. "Values and Shortcomings of Small Reserves." *BioScience* 45 (2): 80–88. https://doi.org/10.2307/1312609.

Shanebeck, Kyle M., Anne A. Besson, Clement Lagrue, and Stephanie J. Green. 2022. "The Energetic Costs of Sub-lethal Helminth Parasites in Mammals: A Meta-analysis." *Biological Reviews* 97 (5): 1886–1907. https://doi.org/10.1111/brv.12867.

Shapiro, Arthur M. 1993. "Thoughts on 'Thoughts on a Review of a Critique for Ecology.'" *Bulletin of the Ecological Society of America* 74 (2): 177–79.

"Share of the Population with Tertiary Education." n.d. Our World in Data. Accessed October 24, 2024. https://ourworldindata.org/grapher/share-of-the-population-with-completed-tertiary-education.

Shaw, Allison K., Chiara Accolla, Jeremy M. Chacón, Taryn L. Mueller, Maxime Vaugeois, Ya Yang, Nitin Sekar, and Daniel E. Stanton. 2021. "Differential Retention Contributes to Racial/Ethnic Disparity in U.S. Academia." *PLOS ONE* 16 (December): e0259710. https://doi.org/10.1371/journal.pone.025971010.1002/essoar.10506980.1.

Shaw, Allison K., and Daniel E. Stanton. 2012. "Leaks in the Pipeline: Separating Demographic Inertia from Ongoing Gender Differences in Academia." *Proceedings of the Royal Society B: Biological Sciences* 279 (1743): 3736–41. https://doi.org/10.1098/rspb.2012.0822.

Shea, Christopher. 1998. "Tribal Skirmishes in Anthropology." *Chronicle of Higher Education*. September 11. https://www.chronicle.com/article/tribal-skirmishes-in-anthropology/.

Shrader-Frechette, Kristin S., and Earl D. McCoy. 1993. *Method in Ecology: Strategies for Conservation*. Cambridge University Press. https://doi.org/10.1017/CBO9780511623394.

Shurin, Jonathan B. 2000. "Dispersal Limitation, Invasion Resistance, and the Structure of Pond Zooplankton Communities." *Ecology* 81 (11): 3074–86. https://doi.org/10.1890/0012-9658(2000)081[3074:DLIRAT]2.0.CO;2.

Shurin, Jonathan B., Elizabeth T. Borer, Eric W. Seabloom, Kurt Anderson, Carol A. Blanchette, Bernardo Broitman, Scott D. Cooper, and Benjamin S. Halpern. 2002. "A Cross-ecosystem Comparison of the Strength of Trophic Cascades." *Ecology Letters* 5 (6): 785–91. https://doi.org/10.1046/j.1461-0248.2002.00381.x.

Shurin, Jonathan B., John E. Havel, Mathew A. Leibold, and Bernadette Pinel-Alloul. 2000. "Local and Regional Zooplankton Species Richness: A

Scale-Independent Test for Saturation." *Ecology* 81 (11): 3062–73. https://doi.org/10.1890/0012-9658(2000)081[3062:LARZSR]2.0.CO;2.

Šigut, Martin, Hana Šigutová, Petr Pyszko, Aleš Dolný, Michaela Drozdová, and Pavel Drozd. 2017. "Avoiding Erroneous Citations in Ecological Research: Read before You Apply." *Oikos* 126 (11): 1523–32. https://doi.org/10.1111/oik.04400.

Silberzahn, R., E. L. Uhlmann, D. P. Martin, P. Anselmi, F. Aust, E. Awtrey, Š. Bahník, et al. 2018. "Many Analysts, One Data Set: Making Transparent How Variations in Analytic Choices Affect Results." *Advances in Methods and Practices in Psychological Science* 1 (3): 337–56. https://doi.org/10.1177/2515245917747646.

Silver, Nate. 2015. *The Signal and the Noise: Why So Many Predictions Fail—but Some Don't*. Illus. ed. Penguin Books.

Simberloff, Daniel. 1981. "The Sick Science of Ecology: Symptoms, Diagnosis, and Prescription." *Eidema* 1:49–54.

———. 2004. "Community Ecology: Is It Time to Move On? (An American Society of Naturalists Presidential Address)." *American Naturalist* 163 (6): 787–99. https://doi.org/10.1086/420777.

Simons, Andrew M. 2011. "Modes of Response to Environmental Change and the Elusive Empirical Evidence for Bet Hedging." *Proceedings of the Royal Society B: Biological Sciences* 278 (1712): 1601–9. https://doi.org/10.1098/rspb.2011.0176.

Smith, Neil. 1987. "'Academic War Over the Field of Geography': The Elimination of Geography at Harvard, 1947–1951." *Annals of the Association of American Geographers* 77 (2): 155–72. https://doi.org/10.1111/j.1467-8306.1987.tb00151.x.

Smith, Richard. 2006. "Peer Review: A Flawed Process at the Heart of Science and Journals." *Journal of the Royal Society of Medicine* 99 (4): 178–82.

Smith, Tyler W., and Jeremy T. Lundholm. 2010. "Variation Partitioning as a Tool to Distinguish between Niche and Neutral Processes." *Ecography* 33 (4): 648–55. https://doi.org/10.1111/j.1600-0587.2009.06105.x.

Smith, Val H., Bryan L. Foster, James P. Grover, Robert D. Holt, Mathew A. Leibold, and Frank deNoyelles. 2005. "Phytoplankton Species Richness Scales Consistently from Laboratory Microcosms to the World's Oceans." *Proceedings of the National Academy of Sciences* 102 (12): 4393–96. https://doi.org/10.1073/pnas.0500094102.

Smolin, Lee. 2007. *The Trouble with Physics: The Rise of String Theory, the Fall of a Science, and What Comes Next*. Illus. ed. Mariner Books.

Soininen, Janne. 2014. "A Quantitative Analysis of Species Sorting across Organisms and Ecosystems." *Ecology* 95 (12): 3284–92. https://doi.org/10.1890/13-2228.1.

Soto, Ismael, Paride Balzani, Laís Carneiro, Ross N. Cuthbert, Rafael Macêdo, Ali Serhan Tarkan, Danish A. Ahmed, et al. 2024. "Taming the Terminological Tempest in Invasion Science." *Biological Reviews* 99 (4): 1357–90. https://doi.org/10.1111/brv.13071.

Sprugel, Doug, ed. 2016a. "Jean L. Richardson." Ecological Society of America, July 23. https://www.esa.org/history/2016/07/richardson-jean/.

———. 2016b. "Steward Pickett." Ecological Society of America, July 22. https://www.esa.org/history/2016/07/pickett-steward/.

Srivastava, Diane S. 1999. "Using Local–Regional Richness Plots to Test for Species Saturation: Pitfalls and Potentials." *Journal of Animal Ecology* 68 (1): 1–16. https://doi.org/10.1046/j.1365-2656.1999.00266.x.

Srivastava, Diane S., Joana Bernardino, Ana Teresa Marques, António Proença-Ferreira, Ana Filipa Filipe, Luís Borda-de-Água, and João Gameiro. 2024. "Editors Are Biased Too: An Extension of Fox et al. (2023)'s Analysis Makes the Case for Triple-Blind Review." *Functional Ecology* 38 (2): 278–83. https://doi.org/10.1111/1365-2435.14483.

Staples, Timothy L., John M. Dwyer, Claire E. Wainwright, and Margaret M. Mayfield. 2019. "Applied Ecological Research Is on the Rise but Connectivity Barriers Persist between Four Major Subfields." *Journal of Applied Ecology* 56 (6): 1492–98. https://doi.org/10.1111/1365-2664.13373.

Stearns, Stephen C. 2000. "Life History Evolution: Successes, Limitations, and Prospects." *Naturwissenschaften* 87 (11): 476–86. https://doi.org/10.1007/s001140050763.

Steenweg, Robin, Mark Hebblewhite, Roland Kays, Jorge Ahumada, Jason T Fisher, Cole Burton, Susan E Townsend, et al. 2017. "Scaling-Up Camera Traps: Monitoring the Planet's Biodiversity with Networks of Remote Sensors." *Frontiers in Ecology and the Environment* 15 (1): 26–34. https://doi.org/10.1002/fee.1448.

Stephens, David W., and John R. Krebs. 1987. *Foraging Theory*. Princeton University Press.

Stocks, Gabriela, Lisa Seales, Franklin Paniagua, Erin Maehr, and Emilio M. Bruna. 2008. "The Geographical and Institutional Distribution of Ecological Research in the Tropics." *Biotropica* 40 (4): 397–404. https://doi.org/10.1111/j.1744-7429.2007.00393.x.

Stomp, Maayke, Jef Huisman, Lajos Vörös, Frances R. Pick, Maria Laamanen, Thomas Haverkamp, and Lucas J. Stal. 2007. "Colourful Coexistence of Red and Green Picocyanobacteria in Lakes and Seas." *Ecology Letters* 10 (4): 290–98. https://doi.org/10.1111/j.1461-0248.2007.01026.x.

Storch, David. 2016. "The Theory of the Nested Species–Area Relationship: Geometric Foundations of Biodiversity Scaling." *Journal of Vegetation Science* 27 (5): 880–91. https://doi.org/10.1111/jvs.12428.

Stouffer, D. B., J. Camacho, R. Guimerà, C. A. Ng, and L. A. Nunes Amaral. 2005. "Quantitative Patterns in the Structure of Model and Empirical Food Webs." *Ecology* 86 (5): 1301–11. https://doi.org/10.1890/04-0957.

Strang, Alexander G., Karen C. Abbott, and Peter J. Thomas. 2019. "How to Avoid an Extinction Time Paradox." *Theoretical Ecology* 12 (4): 467–87. https://doi.org/10.1007/s12080-019-0416-5.

Strauss, Alexander T., Sarah E. Hobbie, Peter B. Reich, Eric W. Seabloom, and Elizabeth T. Borer. 2024. "The Effect of Diversity on Disease Reverses from Dilution to Amplification in a 22-Year Biodiversity × N × CO_2 Experiment." *Scientific Reports* 14 (1): 10938. https://doi.org/10.1038/s41598-024-60725-z.

Strogatz, Steven H. 2004. *Sync: The Emerging Science of Spontaneous Order*. Grand Central Publishing.

Stroud, James T., Michael R. Bush, Mark C. Ladd, Robert J. Nowicki, Andrew A. Shantz, and Jennifer Sweatman. 2015. "Is a Community Still a Community? Reviewing Definitions of Key Terms in Community Ecology." *Ecology and Evolution* 5 (21): 4757. https://doi.org/10.1002/ece3.1651.

Stuart, Yoel E., and Jonathan B. Losos. 2013. "Ecological Character Displacement: Glass Half Full or Half Empty?" *Trends in Ecology & Evolution* 28 (7): 402–8. https://doi.org/10.1016/j.tree.2013.02.014.

Swain, Douglas P., and Ghislain A. Chouinard. 2008. "Viabilité de la population de morue du sud du Golfe du Saint-Laurent." Research Document 2008/018. Canadian Science Advisory Secretariat. Fisheries and Oceans Canada.

Szava-Kovats, Robert C., Argo Ronk, and Meelis Pärtel. 2013. "Pattern without Bias: Local–Regional Richness Relationship Revisited." *Ecology* 94 (9): 1986–92. https://doi.org/10.1890/13-0244.1.

Takacs, David. 1996. *The Idea of Biodiversity: Philosophies of Paradise*. Johns Hopkins University Press.

Tatsioni, Athina, Nikolaos G. Bonitsis, and John P. A. Ioannidis. 2007. "Persistence of Contradicted Claims in the Literature." *JAMA* 298 (21): 2517–26. https://doi.org/10.1001/jama.298.21.2517.

Taylor, L. R. 1961. "Aggregation, Variance and the Mean." *Nature* 189 (4766): 732–35. https://doi.org/10.1038/189732a0.

Taylor, L. R., and I. P. Woiwod. 1982. "Comparative Synoptic Dynamics. I. Relationships between Inter- and Intra-specific Spatial and Temporal Variance/Mean Population Parameters." *Journal of Animal Ecology* 51 (3): 879–906. https://doi.org/10.2307/4012.

Terborgh, John W., and John Faaborg. 1980. "Saturation of Bird Communities in the West Indies." *American Naturalist* 116 (2): 178–95. https://doi.org/10.1086/283621.

"Texas Sharpshooter Fallacy." 2024. Wikipedia. Accessed June 2024. https://en.wikipedia.org/w/index.php?title=Texas_sharpshooter_fallacy&oldid=1248752672.

Thuiller, Wilfried. 2024. "Ecological Niche Modelling." *Current Biology* 34 (6): R225–29. https://doi.org/10.1016/j.cub.2024.02.018.

Tilman, David. 1982. *Resource Competition and Community Structure*. Monographs in Population Biology 17. Princeton University Press.

Tilman, David, Forest Isbell, and Jane M. Cowles. 2014. "Biodiversity and Ecosystem Functioning." *Annual Review of Ecology, Evolution, and Systematics* 45:471–93. https://doi.org/10.1146/annurev-ecolsys-120213-091917.

Tilman, David, Clarence L. Lehman, and Charles E. Bristow. 1998. "Diversity-Stability Relationships: Statistical Inevitability or Ecological Consequence?" *American Naturalist* 151 (3): 277–82. https://doi.org/10.1086/286118.

Todd, Peter A., James R. Guest, Junxiu Lu, and Loke Ming Chou. 2010. "One in Four Citations in Marine Biology Papers Is Inappropriate." *Marine Ecology Progress Series* 408 (June): 299–303. https://doi.org/10.3354/meps08587.

Todd, Peter A., Darren C. J. Yeo, Daiqin Li, and Richard J. Ladle. 2007. "Citing Practices in Ecology: Can We Believe Our Own Words?" *Oikos* 116 (9): 1599–1601. https://doi.org/10.1111/j.0030-1299.2007.15992.x.

Tokeshi, M. 1993. "Species Abundance Patterns and Community Structure." In *Advances in Ecological Research*, vol. 24, edited by M. Begon and A. H. Fitter, 111–86. Academic Press. https://doi.org/10.1016/S0065-2504(08)60042-2.

Touchon, Justin C., and Michael W. McCoy. 2016. "The Mismatch between Current Statistical Practice and Doctoral Training in Ecology." *Ecosphere* 7 (8): e01394. https://doi.org/10.1002/ecs2.1394.

Tredennick, Andrew T., Giles Hooker, Stephen P. Ellner, and Peter B. Adler. 2021. "A Practical Guide to Selecting Models for Exploration, Inference, and Prediction in Ecology." *Ecology* 102 (6): e03336. https://doi.org/10.1002/ecy.3336.

Trimble, Morgan J., and Rudi J. van Aarde. 2012. "Geographical and Taxonomic Biases in Research on Biodiversity in Human-Modified Landscapes." *Ecosphere* 3 (12): art119. https://doi.org/10.1890/ES12-00299.1.

Tuomisto, Hanna. 2010. "A Consistent Terminology for Quantifying Species Diversity? Yes, It Does Exist." *Oecologia* 164 (4): 853–60. https://doi.org/10.1007/s00442-010-1812-0.

Turchin, Peter. 2001. "Does Population Ecology Have General Laws?" *Oikos* 94 (1): 17–26.

Tyre, Andrew J., Hugh P. Possingham, and David B. Lindenmayer. 2001. "Inferring Process from Pattern: Can Territory Occupancy Provide Information about Life History Parameters?" *Ecological Applications* 11 (6): 1722–37. https://doi.org/10.1890/1051-0761(2001)011[1722:IPFPCT]2.0.CO;2.

Ulrich, Werner, Franck Jabot, and Nicolas J. Gotelli. 2017. "Competitive Interactions Change the Pattern of Species Co-occurrences under Neutral Dispersal." *Oikos* 126 (1): 91–100. https://doi.org/10.1111/oik.03392.

Van Valen, Leigh, and Frank A. Pitelka. 1974. "Intellectual Censorship in Ecology." *Ecology* 55 (5): 926–26. https://doi.org/10.2307/1940345.

Vasseur, David A., and Jeremy W. Fox. 2007. "Environmental Fluctuations Can Stabilize Food Web Dynamics by Increasing Synchrony." *Ecology Letters* 10 (11): 1066–74. https://doi.org/10.1111/j.1461-0248.2007.01099.x.

———. 2009. "Phase-Locking and Environmental Fluctuations Generate Synchrony in a Predator-Prey Community." *Nature* 460 (7258): 1007–10. https://doi.org/10.1038/nature08208.

Vellend, Mark. 2016. *The Theory of Ecological Communities*. Monographs in Population Biology 57. Princeton University Press.

Vellend, Mark, Lander Baeten, Isla H. Myers-Smith, Sarah C. Elmendorf, Robin Beauséjour, Carissa D. Brown, Pieter De Frenne, Kris Verheyen, and Sonja Wipf. 2013. "Global Meta-analysis Reveals No Net Change in Local-Scale Plant Biodiversity Over Time." *Proceedings of the National Academy of Sciences* 110 (48): 19456–59. https://doi.org/10.1073/pnas.1312779110.

Vellend, Mark, Maria Dornelas, Lander Baeten, Robin Beauséjour, Carissa D. Brown, Pieter De Frenne, Sarah C. Elmendorf, et al. 2017. "Estimates of Local Biodiversity Change Over Time Stand Up to Scrutiny." *Ecology* 98 (2): 583–90. https://doi.org/10.1002/ecy.1660.

Waldron, Anthony, Daniel C. Miller, Dave Redding, Arne Mooers, Tyler S. Kuhn, Nate Nibbelink, J. Timmons Roberts, Joseph A. Tobias, and John L. Gittleman. 2017. "Reductions in Global Biodiversity Loss Predicted from Conservation Spending." *Nature* 551 (7680): 364–67. https://doi.org/10.1038/nature24295.

Wan, Lingfan, Hao Cheng, Yuqing Liu, Yu Shen, Guohua Liu, and Xukun Su. 2023. "Global Meta-analysis Reveals Differential Effects of Microplastics on Soil Ecosystem." *Science of the Total Environment* 867:161403. https://doi.org/10.1016/j.scitotenv.2023.161403.

Wang, Shaopeng, Forest Isbell, Wanlu Deng, Pubin Hong, Laura E. Dee, Patrick Thompson, and Michel Loreau. 2021. "How Complementarity and Selection Affect the Relationship between Ecosystem Functioning and Stability." *Ecology* 102 (6): e03347. https://doi.org/10.1002/ecy.3347.

Wang, Yongfan, Marc W. Cadotte, Yuxin Chen, Lauchlan H. Fraser, Yuhua Zhang, Fengmin Huang, Shan Luo, Nayun Shi, and Michel Loreau. 2019. "Global Evidence of Positive Biodiversity Effects on Spatial Ecosystem Stability in Natural Grasslands." *Nature Communications* 10 (1): 3207. https://doi.org/10.1038/s41467-019-11191-z.

Ward-Fear, Georgia, Balanggarra Rangers, David Pearson, Melissa Bruton, and Rick Shine. 2019. "Sharper Eyes See Shyer Lizards: Collaboration with Indigenous Peoples Can Alter the Outcomes of Conservation Research." *Conservation Letters* 12 (4): e12643. https://doi.org/10.1111/conl.12643.

Warren, Dan. L., Marcel Cardillo, Dan F. Rosauer, and Daniel I. Bolnick. 2014. "Mistaking Geography for Biology: Inferring Processes from Species Distributions." *Trends in Ecology & Evolution* 29 (10): 572–80. https://doi.org/10.1016/j.tree.2014.08.003.

Warren, P. H., and J. H. Lawton. 1987. "Invertebrate Predator-Prey Body Size Relationships: An Explanation for Upper Triangular Food Webs and Patterns in Food Web Structure?" *Oecologia* 74 (2): 231–35. https://doi.org/10.1007/BF00379364.

Warton, David I., and Francis K. C. Hui. 2011. "The Arcsine Is Asinine: The Analysis of Proportions in Ecology." *Ecology* 92 (1): 3–10. https://doi.org/10.1890/10-0340.1.

Waters, C. Kenneth. 2013. "Shifting Attention from Theory to Practice in Philosophy of Biology." Preprint, PhilSci Archive, September 14. https://philsci-archive.pitt.edu/10009/.

Weisberg, Michael. 2006. "Robustness Analysis." *Philosophy of Science* 73 (5): 730–42. https://doi.org/10.1086/518628.

Welden, C. W., and W. L. Slauson. 1986. "The Intensity of Competition versus Its Importance: An Overlooked Distinction and Some Implications." *Quarterly Review of Biology* 61 (1): 23–44. https://doi.org/10.1086/414724.

West, G. B., J. H. Brown, and B. J. Enquist. 1997. "A General Model for the Origin of Allometric Scaling Laws in Biology." *Science* 276 (5309): 122–26. https://doi.org/10.1126/science.276.5309.122.

West, Jevin D., Jennifer Jacquet, Molly M. King, Shelley J. Correll, and Carl T. Bergstrom. 2013. "The Role of Gender in Scholarly Authorship." *PLOS ONE* 8 (7): e66212. https://doi.org/10.1371/journal.pone.0066212.

White, Craig R., Dustin J. Marshall, Steven L. Chown, Susana Clusella-Trullas, Steven J. Portugal, Craig E. Franklin, and Frank Seebacher. 2021. "Geographical Bias in Physiological Data Limits Predictions of Global Change Impacts." *Functional Ecology* 35 (7): 1572–78. https://doi.org/10.1111/1365-2435.13807.

Whittaker, R. H. 1960. "Vegetation of the Siskiyou Mountains, Oregon and California." *Ecological Monographs* 30 (3): 279–338. https://doi.org/10.2307/1943563.

Wiens, John J. 2023. "How Many Species Are There on Earth? Progress and Problems." *PLOS Biology* 21 (11): e3002388. https://doi.org/10.1371/journal.pbio.3002388.

Wiersma, Yolanda F. 2022. "A Review of Landscape Ecology Experiments to Understand Ecological Processes." *Ecological Processes* 11 (1): 57. https://doi.org/10.1186/s13717-022-00401-0.

Williams, Richard J., and Neo D. Martinez. 2000. "Simple Rules Yield Complex Food Webs." *Nature* 404 (6774): 180–83. https://doi.org/10.1038/35004572.

Williamson, Donald I. 2009. "Caterpillars Evolved from Onychophorans by Hybridogenesis." *Proceedings of the National Academy of Sciences* 106 (47): 19901–5. https://doi.org/10.1073/pnas.0908357106.

Willig, M. R., D. M. Kaufman, and R. D. Stevens. 2003. "Latitudinal Gradients of Biodiversity: Pattern, Process, Scale, and Synthesis." *Annual Review of Ecology, Evolution, and Systematics* 34:273–309. https://doi.org/10.1146/annurev.ecolsys.34.012103.144032.

Willmer, Julian Nicholas G., Thomas Püttker, and Jayme Augusto Prevedello. 2022. "Global Impacts of Edge Effects on Species Richness." *Biological Conservation* 272 (August): 109654. https://doi.org/10.1016/j.biocon.2022.109654.

Wilson, William G., and Per Lundberg. 2004. "Biodiversity and the Lotka–Volterra Theory of Species Interactions: Open Systems and the Distribution of Logarithmic Densities." *Proceedings of the Royal Society B: Biological Sciences* 271 (1551): 1977–84. https://doi.org/10.1098/rspb.2004.2809.

Wilson, William G., P. Lundberg, D. P. Vázquez, J. B. Shurin, M. D. Smith, W. Langford, K. L. Gross, and G. G. Mittelbach. 2003. "Biodiversity and Species Interactions: Extending Lotka–Volterra Community Theory." *Ecology Letters* 6 (10): 944–52. https://doi.org/10.1046/j.1461-0248.2003.00521.x.

Wimsatt, William C. 2007. *Re-engineering Philosophy for Limited Beings: Piecewise Approximations to Reality*. Harvard University Press.

Wittgenstein, Ludwig. 2010. *Philosophical Investigations*. 4th ed. Edited by P. M. S. Hacker and Joachim Schulte. Wiley-Blackwell.

Woit, Peter. 2006. *Not Even Wrong: The Failure of String Theory and the Search for Unity in Physical Law*. 1st ed. Basic Books.

Woods, Natasha N., Zakiya H. Leggett, and Maria N. Miriti. 2023. "The Intersections of Identity and Persistence for Retention in Ecology and Environmental Biology with Personal Narratives from Black Women." *Journal of Geoscience Education* 71 (3): 332–43. https://doi.org/10.1080/10899995.2022.2154935.

Wootton, J. T., and C. A. Pfister. 1998. "The Motivation of and Context for Experiments in Ecology." In *Experimental Ecology: Issues and Perspectives*, edited by William J. Resetarits and Joseph Bernardo, 350–69. Oxford University Press.

Wortley, Liana, Jean-Marc Hero, and Michael Howes. 2013. "Evaluating Ecological Restoration Success: A Review of the Literature." *Restoration Ecology* 21 (5): 537–43. https://doi.org/10.1111/rec.12028.

Wright, Alexandra J., Kathryn E. Barry, Christopher J. Lortie, and Ragan M. Callaway. 2021. "Biodiversity and Ecosystem Functioning: Have Our Experiments and Indices Been Underestimating the Role of Facilitation?" *Journal of Ecology* 109 (5): 1962–68. https://doi.org/10.1111/1365-2745.13665.

Wright, Jacqueline D., Jeffrey P. Hughes, Yechiam Ostchega, Sung Sug Yoon, and Tatiana Nwankwo. 2011. "Mean Systolic and Diastolic Blood Pressure in Adults Aged 18 and Over in the United States, 2001–2008." National Health Statistics Reports 35. US Department of Health and Human Services, Centers for Disease Control and Prevention, National Center for Health Statistics, March 25.

Xiao, Xiao, Kenneth J. Locey, and Ethan P. White. 2015. "A Process-Independent Explanation for the General Form of Taylor's Law." *American Naturalist* 186 (2): E51–60. https://doi.org/10.1086/682050.

Yaari, Gur, Yossi Ben-Zion, Nadav M. Shnerb, and David A. Vasseur. 2012. "Consistent Scaling of Persistence Time in Metapopulations." *Ecology* 93 (5): 1214–27. https://doi.org/10.1890/11-1077.1.

Yachi, Shigeo, and Michel Loreau. 1999. "Biodiversity and Ecosystem Productivity in a Fluctuating Environment: The Insurance Hypothesis." *Proceedings of the National Academy of Sciences* 96 (4): 1463–68. https://doi.org/10.1073/pnas.96.4.1463.

Yang, Yefeng, Helmut Hillebrand, Malgorzata Lagisz, Ian Cleasby, and Shinichi Nakagawa. 2022. "Low Statistical Power and Overestimated Anthropogenic Impacts, Exacerbated by Publication Bias, Dominate Field Studies in Global Change Biology." *Global Change Biology* 28 (3): 969–89. https://doi.org/10.1111/gcb.15972.

Yang, Yefeng, Alfredo Sánchez-Tójar, Rose E. O'Dea, Daniel W. A. Noble, Julia Koricheva, Michael D. Jennions, Timothy H. Parker, Malgorzata Lagisz, and Shinichi Nakagawa. 2023. "Publication Bias Impacts on Effect Size, Statistical Power, and Magnitude (Type M) and Sign (Type S) Errors in Ecology and Evolutionary Biology." *BMC Biology* 21 (1): 71. https://doi.org/10.1186/s12915-022-01485-y.

Yoder, Jeremy B., and Allison Mattheis. 2016. "Queer in STEM: Workplace Experiences Reported in a National Survey of LGBTQA Individuals in Science, Technology, Engineering, and Mathematics Careers." *Journal of Homosexuality* 63 (1): 1–27. https://doi.org/10.1080/00918369.2015.1078632.

Zapata, Fernando A., K. J. Gaston, and S. L. Chown. 2003. "Mid-domain Models of Species Richness Gradients: Assumptions, Methods and Evidence." *Journal of Animal Ecology* 72 (4): 677–90. https://doi.org/10.1046/j.1365-2656.2003.00741.x.

———. 2005. "The Mid-domain Effect Revisited." *American Naturalist* 166 (5): E144–48; discussion on E149–54. https://doi.org/10.1086/491685.

Zeder, Melinda A. 2018. "Why Evolutionary Biology Needs Anthropology: Evaluating Core Assumptions of the Extended Evolutionary Synthesis." *Evolutionary Anthropology: Issues, News, and Reviews* 27 (6): 267–84. https://doi.org/10.1002/evan.21747.

Zettlemoyer, Meredith A., Karina M. Cortijo-Robles, Nicholas Srodes, and Sarah E. Johnson. 2023. "What Are We Reading? Hot Topics and Authorship in Ecology Literature across Decades." *Bulletin of the Ecological Society of America* 104 (1): e02025. https://doi.org/10.1002/bes2.2025.

Index

Page numbers in italics refer to figures.

Abbott, Karen, 128–29
abduction, 100. *See also* inference to the best explanation
AIC (Akaike information criterion), 19, 32, 39
Allee effect, 128–29
allometric scaling, 113–14, 116
alpha diversity. *See* diversity
alternate attractors, 95–96
analogies, fruitful, 181–85
Anderson, Sean C., 16, 19
anoles, 146–47, 158–59
anthropology, 205–6
Archilochus (Greek poet), 215
Ashworth, Scott, 201–4
astronomy, 93, 101–3
attractors: alternate, 95–96; statistical, 179, 185–87
author selectivity, 87–92
Avgar, Tal, 30

behavior, 3, 32, 38, 53, 74, 85, *86*, 135, 147–50, *150*, 153, 172, 183; of mathematical models, 127–28, 142, 176; oscillatory, 112
Berlin, Isaiah, 215–16
Bernardo, Joseph, xiii
Berry, Christopher, 201–4
Besnard, Aurélien, 157
beta diversity. *See* diversity
bet hedging, 73
Betini, Gustavo, 30
Betts, Matthew G., 30
biodiversity, *16*, 18, 37, 46, 50, 148; coinage, 190–91; controversy regarding declines, *77*, 151–53; habitat fragmentation effects, 75, 88–91, 148, 156, 193; vague definitions, 190–91. *See also* biodiversity-ecosystem function; diversity; local-regional richness relationships
biodiversity-ecosystem function (BEF), 46, 49–51, 55, 66, 69, 71–72, 82, 92, 103, 110, 145, 151, 189. *See also* biodiversity
biological invasions, 16, 18. *See also* invasive species
black chiton, 54–55
body size, 53, 96, 113–14, *115*, 132–33, 136, 184
Bolker, Ben, 12–13
Boltzmann equation, 103, 105
bovine tuberculosis (TB), 62–65

Bowler, Diana, 32
broken-stick models, 182
Brown, James, 161
buffalo, 61–65

Cambrian explosion, 153
Carpenter, Steve, 10, 58
case studies, 3, 42, 107, 146, 161, 180–81, 186–87, 205
central limit theorem, 178
character displacement, 73
chiton, 54–55
Chitty, Dennis, xii
citations: of applied papers, 18; like a game of telephone, 147–48; networks of, 185; and peer review history, 79, *80*; and selective presentation of results, 90; of statistical software, *39*
citizen science, 27, 29, 53
Clark, Timothy D., 149–50, 156
Clements, Jeff C., 149–50, *150*
climate change, 1, 16, 18, 25–26, 38, 41, 74, 145, 175, 211, 218
coexistence theory, 132, 179–80, 191, 194
collaboration, 22, 25, 29, 40
competition: checkerboard distributions, 92, 104–5; compensatory dynamics, 124–25, *126*; Hutchinson's ratio, 96–97; importance vs. intensity, 192–93; randomized null models, 104–5
complementarity effects: in biodiversity-ecosystem function experiments, 49–51; causes of, 49–50; contrast to selection effects, 49–50, 72; definition of, 49–50; negative, 189; in scientific research, 51, 110, 218
continental drift, 99–100
controversies, 74–75, *76*, *77*, 100, 106, 113, 117, 210
Cooper, Gregory, 1
coordinated distributed experiments, 211–14, 219
Costello, Laura, 37
coupled oscillators, 181, 183
Craven, Dylan, 18
A Critique for Ecology (Peters), 210–11

Daphnia spp., 72–73, 129–30
Darwin, Charles, 45, 179, 206–7
data sharing, 21–23, 40
deduction, 121
Diamond, Jared, 104
Didinium nasutum, 134–35
disease, 25–26, 60–61, 63, 94, 98, 105–6, 116. *See also* bovine tuberculosis; buffalo; *Daphnia* spp.
disturbances, environmental, 124
diversification, 18–19, 21–25
diversity: alpha, beta, and gamma, 131, 133, 191; of ecologists, 1, 11, 15, 218; partitioning, 131, 133, 191. *See also* biodiversity; diversification; local-regional richness relationships; research approaches; research goals
Dixson, Danielle, 148–50
DNA, 53, 207
Dochtermann, Ned A., 74
Dornelas, Maria, 151–53, 159
Downes, Barbara J., 74
Dresow, Max, 153–55
Duffy, Meghan, 129–31
Dunne, Jennifer A., 185
Dynamic Ecology blog, 1–2, 42, *43*, 44, 74, 180, 213, 215–16

ecosystem function. *See* biodiversity-ecosystem function
ecosystem health, 192
ecosystem integrity. *See* ecosystem health
Ediacaran organisms, 153–55

Efron, Bradley, 56
Egerton, Frank, xi
Egger's regression, 83–85, *87*
Eichhorn, Marcus, 13–15
Ellner, Stephen P., 139–40
Ellstrand, Norman, 156–57
Elton, Charles, 111–12, 120
environmental stochasticity, 95–96, 142
evolutionary biology, 156, 182, 194; analogy to community ecology, 187–88; as a diverse and adaptive scholarly field, 206–8
Ezenwa, Vanessa, 61–65

fabricated results, 150
facilitation, 49, 52, 55–57, 59; accidental vs. intentional, 55–56
Fahrig, Lenore, 88–90, 92, 192–93
field experiments. *See* research approaches
field observations. *See* research approaches
file drawer problem, 67, 82–85
Fischer, Joern, 192
foot-and-mouth, 26, 116. *See also* disease
Ford, Adam T., 74
forecasting. *See* research goals
Fourcade, Yoan, 157
Fournier, Auriel, 11
Fox, Charles, 78–81, *80*
fox and hedgehog metaphor, 214–19
Freckleton, Robert P., 192–93
Fretwell, Stephen D., 157
fruitful analogies, 181–85
Fryxell, John, 30
functional response, 127–28, 134–35
funnel plot, 83–85, *86*. *See also* file drawer problem

gamma diversity. *See* diversity
Gause, G. F., 9
geography, 204–5
Glaessner, Martin, 153–55
Glazier, Douglas S., 116
Glick, Thomas P., xii
global change. *See* climate change
Godfrey-Smith, Peter, xiii
Goheen, Jacob R., 74
Gómez, Pablo, xi–xii
Gould, Elliot, 196–97
Gould, Stephen Jay, 156, 158, 159
Gowers, William Timothy, 200–201
Grogan, Paul, 30, 32
Gurevitch, Jessica, 20, 21

habitat fragmentation, 74, 88–92, 148, 156, 192
Hall, Spencer, 129–31
HARKing, 100–101
Harrison, Gary, 134–35
Harte, John, 161
Hector, Andy, 138
hedgehog. *See* fox and hedgehog metaphor
helminths, 61–63, *64*
Hintzen, Rogier E., 18
Hoeksema, Jason D., 162–66, *165*
hoof-and-mouth. *See* foot-and-mouth
Hui, Francis K. C., 156
humor in scientific communication, 156
Huntington, Henry, 54
Hutchinson, G. Evelyn, 10, 96, 121–22
Hutchinson's ratio, 96–97, 108–9
hypotheses: controversial, 74; about disease dynamics, 62; lack of, 29–32, 72; lines of evidence for, 129; null, 37–38, 168; scientific vs. statistical, 29–30; speculative, 155, 207; testing, 34, 121, 212; working, 119. *See also* HARKing; inference to the best explanation; strong inference

impact factor, 79, *80*
indirect evidence, 55–57, 60
inference to the best explanation, 99–101, 103
inferring process from pattern, 92–108
Intergovernmental Panel on Climate Change (IPCC), 38
invasion biology, 16, 18. *See also* invasive species
invasive species, 68, 90, 168–69. *See also* invasion biology
island biogeography, 170–72
Ives, Tony, 95–96, 161, 181

Jenkins, Stephen H., 74
Jennions, Michael, 37–38
Jolles, Anna, 63–65, *64*
journals, 16, 18, 20–22, *20*, 29–32, *31*, *33*, 35, *39*, 68, 85, 90, 149, 157, 204, 210–11, *211*; new, 17; selectivity of, 70, 76, 78–83. *See also* file drawer problem; peer review; publication bias

Kamath, Ambika, xii, 147, 156, 158, 159
Kareiva, Peter, 161
Kendall, Bruce E., 72
kidney function, 56–57. *See also* indirect evidence; linear regression
Kingsland, Sharon, xi
Klausmeier, Christopher A., 142–43, *143*
Koper, Nicola, 90
Koricheva, Julia, 83
Krebs, Charles, 32, 58

lab experiments. *See* research approaches
laws, scientific, 3, 160–61, 181, *214*
Lawton, John, 161, 187
Leibold, Mathew A., 141–42
Lele, Subhash, xiii
Lemoine, Nathan P., 37, 38
Levins, Richard, 174, 175, 176, 216
Lévy walks, 107
Lewontin, R. C., 156, 158, 159
lichens, coverage claim, 148
limiting similarity, 96, 171
Lindenmayer, David B., 192
linear regression, 19, 56–58, *59*, 83, 115, *140*. *See also* indirect evidence
Linnell, John, 32
local-regional richness relationships, 98, 108
Locey, Kenneth, 138–39, *140*
lognormal distribution: and Hutchinson's ratio, 96–97; of species abundances, 178–79
Loreau, Michel, 138
Losos, Jonathan, 147, 156, 159
Luckinbill, Leo, 134–35

MacArthur, Robert, 2, 10, 93, 96, 115–16, 134–35, 157, 160, 170–73, *171*, 174
macroinvertebrates, 53
many analysts/one dataset, 195–97
marginal value theorem, 171
Martin, Laura, xi
Martinez, Neo D., 135, 185
mathematics, pure, 199–201
May, Robert M., 120
May-Wigner theorem, 171
McCallen, Emily, 18
McCauley, Edward, 72–73
McCoy, Earl, 186–87
McPhail, John Donald, 73
medical diagnosis, 94–99, 103, 105–7
Menzies, Allyson, 11–12
Mesquita, Ethan Bueno de, 201–4
meta-analysis, 2, 19, 20–22, 37, 40, 42, *43*, 53, 83, 85, 149, 162–70, *165*, *166–67*, 180, 187, 212, *214*, 217; heterogeneity in, 84, *87*, 164, 166–70, 218; limitations of, 168–70

metacommunity, 141–42, 180
metapopulation, 89, 141
microcosms: complementary to studies of natural water bodies, 58–60; criticisms of, 10, *43*, 58–60; history of use in ecology, 28; protist, 9, 10, 216. *See also* research approaches
microplastics, 85, *86*
mid-domain effect, 105–6. *See also* Narcissus effect
midges, 95
Milo, Ron, 185
model selection, *16*, 19, 32, 66
Møller, Anders, 37–38
Munday, Philip, 148–50
Murdoch, William W., 173
Murray, Jim, 112
mycorrhizal fungi, 25, 162–65, 168

Narcissus effect, 104, 106. *See also* mid-domain effect
National Center for Ecological Analysis and Synthesis (NCEAS), 40
natural history, 2, 8, 12, 42
networks: metacommunities, 180; of nodes and links, 185–86
neutral models, 141–42
niche, 98, 191, 194; construction, 207; ecological niche modeling, 191, 194
Nilsen, Erlend, 32
normal distribution, 176, 178–79
null models, 103–5, 137–40, *140*
NutNet (Nutrient Network), 212–13

O'Brien, Allyson, 192
observational field studies. *See* research approaches
ocean acidification, 148–50, *150*, 153
O'Connor, Mary I., 74
Ollerton, Jeff, 12
operationalization, 190, 192–93, 195
optimal foraging theory, 171–72
optimal life histories, 172, 174
Osseo-Asare, Abena Dove, xii

Packer, Melina, xii
Paine, Tim, 78–81, *80*
Paramecium spp., 9, 134–35
pattern vs. process, 92–108
Payne-Gaposchkin, Cecilia, 103, 105
peer review, 70, 76–79, 81, 88, 109, 157–58
Peters, Robert, 158, 210–11
physics, 3, 13, 160, 181, 183, 206, 208
phytoplankton, 57–60, 142–43
Pickett, Steward, 11
Pitelka, Frank, 160
plate tectonics, 93, 99
Platt, John, 71
political science, 201–4
population cycles, 72, 95, 134, 173; spatial synchrony of, 175–76, 181, 183
population genetics, 160, 184
power, statistical, 35–37, 59, 217
power law, 58, 107, 113, 139, 184
predator-prey, 72, 115, 127, 134–35, 171
Price, George, 132
Price equation, 132–33, 182
process vs. pattern, 92–108
protist microcosms, 9, 10, 216
publication bias, 83–85, *86*. *See also* file drawer problem
pure mathematics, 199–201

random draws design, 47, 49, 138
range shifts, 175
Rees, Mark, 192–93
regression. *See* linear regression
research approaches, 5–6, 25, 27, 32–35, *33*, 41, 51, 55, 63, 70, 93, 112, 144, 162, 184, 208, 211–13, 219; data synthesis, 28; field experiments, 28; field observations, 28, 33–34; laboratory/microcosm/mesocosm experiments, 28;

research approaches (*cont.*)
mathematical modeling, 28; new approaches, 211; other approaches, 28; proposing new methods, 28; sharing opinions, 28
researcher degrees of freedom, 195. *See also* many analysts/one dataset
research goals, 15, 25, 27–29, 44, 204, 209–10; description, 25; explanation, 26; management, 26; prediction or forecasting, 26, 210–11, 214, 219
Resetarits, William J., xiii
resource exchange, 52, 55. *See also* facilitation
rhetoric, 156, 158
Richardson, Jean, 12
Riva, Federico, 90
robust theorems, 174–76
Roche, Dominique G., 22
Rosen, Walter G., 190
Rosenzweig, Michael L., 113–16, 134–35
Rosenzweig–MacArthur model, 115, 134–35, 171
Rovere, Jacob, 136, *137*
R software, 39–41

Saha equation, 103, 105
Salomon, Anne, 54
sample size, *34*, 35–37, 83, 147, 164, 166, 168
sampling effect, 138
satire in scientific communication, 156–57
Saunders, Manu E., 157
Schluter, Dolph, 73
scientific laws, 3, 160–61, 181, *214*
Secondi, Jean, 157
Seilacher, Adolf, 154–55
selection effects: in biodiversity-ecosystem function experiments, 46–48, 71–72; contrast to complementarity effects, 49–50, 72; definition, 46–48; negative, 69–70, 82–83, 92–94, 189; in scientific research, 69–70, 83, 92–94, 108, 146, 218
Sells, Sarah N., 32
Shrader-Frechette, Kristin, 186–87
Silberzahn, Raphael, 195–96
Simons, Andrew M., 73
Sinclair, George, 45, 47
Slauson, William L., 192
Smith, Val H., 57–60, *59*
spandrels analogy, 156
spatial synchrony. *See* population cycles
species-abundance distribution, 106, 179, 182, 186
species-area curve, 57–60, 114
spectrogram, 103
spectroscopy, 101–2
stability, 120, 171, 191, 193
Staples, Timothy L., 18
statistical attractor, 179, 185–87
statistical power, 35–37, 59, 217
storage effect, 139–40
Strang, Alexander, 128–29
stressors, 36, 74
Strevens, Michael, xii–xiii
strong inference, 71–74
synchrony, spatial. *See* population cycles

Takacs, David, xiii, 190–91
Tanape, Nick, Sr., 54
Taper, Mark, xiii
Taylor's law, 113, 138–39, *140*
Texas sharpshooter fallacy, 100
Thomas, Peter, 128–29
Tower of Babel, 189–91
trophic cascades, 74, 187
tuberculosis. *See* bovine tuberculosis
Turchin, Peter, 161
two cultures of mathematics, 200–201
Type M errors, 38

unifying theoretical frameworks, 179–80, 184–85, 187

vagueness, 5, 32, *43*, 119, 162, 180–81, 186, 190–94
Van Valen, Leigh, 160
Vasseur, David, 124–25, *126*, 176
Vellend, Mark, 151–53, 159, 185, 188

Warton, David I., 156
Waters, C. Kenneth, 194
Watkinson, Andrew R., 192–93
Wegener, Alfred, 99, 100
Welden, Charles W., 192
White, Ethan, 138–39, *140*
Whittaker, Robert, 191
Williams, Richard J., 135, 185
Wilson, Edward O., 170–73, *171*
Wright-Fisher model, 184

Xiao, Xiao, 138–39, *140*

Yang, Yefeng, 37, 85, 87

zombie ideas, 224
zooplankton, 58, 72, 129, 136. *See also* *Daphnia* spp.

www.ingramcontent.com/pod-product-compliance
Lightning Source LLC
Chambersburg PA
CBHW032119020426
42334CB00016B/1012